WELDING PROCESSES AND PRACTICES WORKBOOK

WELDING PROCESSES AND PRACTICES WORKBOOK

Eugene G. Hornberger

August F. Manz

John Wiley & Sons

New York • Chichester • Brisbane • Toronto • Singapore

Copyright © 1988, by John Wiley & Sons, Inc.

All rights reserved. Published simultaneously in Canada.

Reproduction or translation of any part of
this work beyond that permitted by Sections
107 and 108 of the 1976 United States Copyright
Act without the permission of the copyright
owner is unlawful. Requests for permission
or further information should be addressed to
the Permissions Department, John Wiley & Sons.

ISBN 0-471-81668-x

Printed in the United States of America

10 9 8 7 6 5 4 3 2 1

Preface

The training of welders requires a great deal of individualized instruction. Often classes are large and stretch the instructor's ability to devote enough attention to each student.

Welding instructors need instructional material that will reduce the amount of one-on-one time required for each student. This workbook provides students with reference material so that they will not need to approach the instructor on items already covered. Thus students can use the reference material for reinforcement of their skills.

The main purpose of this workbook is to present the student with a wide variety of joint types and in a variety of positions, using a selection of the more popular electrodes used today.

The workbook does not purport to replace the instructor; the instructor must always be the ultimate authority and reference for the student.

The angles and manipulations for the electrodes and torches will do the job properly. Should the instructor have other preferences, they would certainly supersede those presented here.

The hands-on Procedures are designed to aid the instructor, and to assist the student in learning the fundamentals of welding. They should also reduce the number of times the student will have to go to the instructor with questions.

The Procedures should also assist the student in learning to follow written instructions as well as to develop a sense of responsibility toward the job.

Most of the questions students may have are covered along with descriptive drawings. In most cases there is further descriptive and explanatory information in the corresponding lesson in the *Welding Processes and Practices* textbook. Each Procedure in the workbook has been conveniently cross-referenced to its exact Section, Lesson, and page number in the textbook. Lessons should first be studied in that text, then the workbook should be used as a guide while performing the task.

Following the precepts of vocational education, it is suggested that the Procedure sheets be used in conjunction with the instructor's lecture and demonstration as supportive material.

To the Student

Before you start, use the proper safety equipment and work efficiently at all times.

Everything you learn here will be helpful in other phases of your training. The good habits you form will stay with you, but so will the bad ones. Do the best you can from the beginning; bad habits are difficult to break.

One of the more difficult things you must learn is to set your welding conditions properly. In each Procedure you will be given a range of welding conditions for the exercise you will be doing. You will have to find the point in that range that produces the best weld for you. This will come through practice.

It is important for a welder to be able to set the correct welding conditions for the material type, size, thickness, or position. All these factors are important to know when setting up the equipment. For this reason the welder must not only be able to set gas pressures, amperages, voltages, or gas flows, they must be able to "read" the puddle and recognize the sound it makes when the conditions are correct. Through practice the welder can recognize when the welding conditions are proper by the appearance of the flame or the sound the arc makes and by the way the puddle reacts.

SAFETY REMINDERS

In addition to the safety reminders in the procedures, you should always remember to:

1. Pick up hot metal with tongs, pliers, or clamps. *Never pick up hot metal with your gloves or bare hands. Gloves used to pick up hot metal will quickly become stiff, brittle, and unusable.*
2. Cool all metal after welding.
3. Remove all welding stubs and wire from the area.
4. Leave the area clean for the next student.
5. Secure the welding equipment in the proper place at the end of the Procedure.

GENERAL MATERIAL AND EQUIPMENT

The general material and equipment you will need are listed at the beginning of each section. Additional materials needed for specific Procedures are designed in the specification boxes within each Procedure.

GENERAL INSTRUCTIONS

Throughout these Procedures you will be using a variety of equipment. When direct current is called for, the electrode charge or polarity will be specified on the technique sheet.

One final note before you start your hands-on training. The Procedures in this book have been written for right-handed welders. Left-handed students should start at the opposite end of the plate and reverse all the directions and angles given.

Welders should learn to work both from left to right and from right to left. In addition, they should learn to weld toward and away from themselves. A proficient welder can weld in any direction.

However, there is something very important to remember before you start: *Namely, your instructor is the expert.* Go to your instructor whenever you have a problem. Do not move on to a new Procedure before the instuctor has demonstrated that Procedure (unless, of course, you have permission to do so).

Contents

OXYFUEL GAS WELDING AND BRAZING — 1

Oxyacetylene Welding (OAW) of Carbon Steel — 3

Procedure 1	Flat Beads Without Filler Metal	5
Procedure 2	Flat Beads with Filler Metal	7
Procedure 3	Flat Outside-Corner Joint	9
Procedure 4	Flat Outside-Corner Joint with Filler Metal	11
Procedure 5	Flat Square-Groove Butt Joint with Filler Metal	13
Procedure 6	Horizontal Outside-Corner Joint	15
Procedure 7	Horizontal Square-Groove Butt Joint	17
Procedure 8	Horizontal Fillet Joint Position	19
Procedure 9	Vertical Outside-Corner Joint, Uphill Weld	21
Procedure 10	Vertical Outside-Corner Joint, Downhill Weld	23
Procedure 11	Vertical Square-Groove Butt Joint	25
Procedure 12	Vertical Fillet Joint	27
Procedure 13	Overhead Outside-Corner Joint	29
Procedure 14	Overhead Square-Groove Butt Joint	31
Procedure 15	Overhead Fillet Joint	33

Oxyacetylene Braze-Welding — 35

Procedure 16	Running Flat Beads	37
Procedure 17	Flat V-Groove Butt Joint	39
Procedure 18	Horizontal T-Joint	41
Procedure 19	Horizontal Lap Joint Fillet	43
Procedure 20	Horizontal V-Groove Butt Joint	45
Procedure 21	Vertical T-Joint Fillet	47
Procedure 22	Vertical V-Groove Butt Joint	49

Oxyacetylene Welding of Pipe — 51

Procedure 23	IG Position, Axis of the Pipe Horizontal	53
Procedure 24	2G Position, Axis of the Pipe Vertical	55
Procedure 25	5G Position, Axis of the Pipe Horizontal	57

SHIELDED METAL ARC WELDING PROCESS — 59

Shielded Metal Arc Welding of Plate — 61

Procedure 26	Striking and Maintaining an Arc	63
Procedure 27	Welding Stringer Beads	65
Procedure 28	Running a Flat Pad of Stringers	67
Procedure 29	T-Joint Fillet, 1F Position, Stringer Beads (E-6010 or E-6011 Electrodes)	69
Procedure 30	T-Joint Fillet, 1F Position, Stringer Beads (E-7014 or E-7024 Electrodes)	71
Procedure 31	Filling Craters at the End of Welds	73
Procedure 32	Square-Groove Butt Joint, 1G Position With Backup Bar, Stringer Beads	75
Procedure 33	Outside-Corner Joint, 1G Position, Stringer and Weave Beads	77
Procedure 34	Running a Horizontal Pad of Stringers	79
Procedure 35	T-Joint Fillet, 2F Position, Multipass Stringers (E-6010 or E-6011 Electrodes)	81
Procedure 36	T-Joint Fillet, 2F Position, Multipass Stringers (E-7024 Electrode)	83
Procedure 37	Single-Bevel Butt Joint, 2G Position with Backup Bar	85
Procedure 38	Open-Root V-Groove, Butt Joint, 2G Position	87
Procedure 39	T-Joint Fillet, 3F Position, Weave Beads, Uphill (E-6010 or E-6011 Electrodes)	89
Procedure 40	T-Joint Fillet, 3F Position, Weave Beads, Uphill (E-7018 Electrode)	91
Procedure 41	T-Joint Fillet, 3F Position, Stringer Beads, Downhill	93
Procedure 42	V-Groove Butt Joint, 3G Position with Backup Bar, Uphill	95
Procedure 43	Open-Root V-Groove Butt Joint, 3G Position, Stringer and Weave Beads	97
Procedure 44	T-Joint Fillet, 4F Position, Stringer Beads (E-6010 or E-6011 Electrodes)	99
Procedure 45	T-Joint Fillet, 4F Position, Stringer Beads (E-7018 or E-7024 Electrodes)	101
Procedure 46	V-Groove Butt Joint, 4G Position with Backup Bar (E-6010 or E-6011 Electrodes)	103
Procedure 47	V-Groove Butt Joint, 4G Position with Backup Bar (E-7018 Electrode)	105
Procedure 48	Open-Root V-Groove Butt Joint, 4G Position	107

Shielded Metal Arc Welding of Sheet Metal — 109

Procedure 49	Welding an Outside-Corner Joint in the Flat or Downhand Position	111
Procedure 50	T-Joint Fillet, 2F Position	113
Procedure 51	Lap Joint Fillet, 2F Position	115
Procedure 52	Outside-Corner Joint, 3G Position (E-6011 Electrode)	117
Procedure 53	Outside-Corner Joint, 3G Position (E-6013 Electrode)	119
Procedure 54	Lap Joint Fillet, 3F Position	121
Procedure 55	Lap Joint Fillet, 4F Position	123
Procedure 56	Outside-Corner Joint, 4G Position	125
Procedure 57	Welding a Six-Sided Box, in the 2G, 3G, and 4G Positions	127

Shielded Metal Arc Welding of Pipe — 129

Procedure 58	Beveling Pipe to Prepare It for Tests	131
Procedure 59	Alignment, Fit-Up, and Tacking Procedure	133
Procedure 60	Welding Pipe in the 1G Position, with Its Axis in the Horizontal Position	135
Procedure 61	Welding Pipe in the 2G Test Position, with Its Axis in the Vertical Position (E-6010 or E-6011 Electrodes)	137
Procedure 62	Welding Pipe in the 2G Test Position, with Its Axis in the Vertical Position (E-6010 or E-6011 and E-7018 Electrodes)	139
Procedure 63	Welding Pipe in the 5G Position Uphill Passes (E-6010 or E-6011 Electrodes)	141
Procedure 64	Welding Pipe in the 5G Position First Pass Downhill (E-6010 or E-6011 Electrodes) Second and Out Uphill (E-7018 Electrodes)	143
Procedure 65	Welding Pipe in the 5G Position Downhill Passes (E-6010 or E-6011 Electrodes)	145
Procedure 66	Welding Pipe in the 6G Position Downhill Passes (E-6010 or E-6011 Electrodes)	147
Procedure 67	Welding Pipe in the 6G Position First Pass Downhill (E-6010 and E-6011 Electrodes), Second and Out Uphill (E-7018 Electrode)	149

GAS TUNGSTEN ARC WELDING PROCESS — 151

Gas Tungsten Arc Welding of Carbon Steel — 153

Procedure 68	Running a Flat Pad of Stringers	155

xii CONTENTS

Procedure 69	Outside-Corner Joint, 1G Position	**157**
Procedure 70	T-Joint Fillet, 1F Position	**159**
Procedure 71	V-Groove Butt Joint, 1G Position	**161**
Procedure 72	Outside-Corner Joint, 2G Position	**163**
Procedure 73	T-Joint Fillet, 2F Position	**165**
Procedure 74	V-Groove Butt Joint, 2G Position	**167**
Procedure 75	Outside-Corner Joint, 3G Position	**169**
Procedure 76	T-Joint Fillet, 3F Position	**171**
Procedure 77	V-Groove Butt Joint, 3G Position	**173**
Procedure 78	Outside-Corner Joint, 4G Position	**175**
Procedure 79	T-Joint Fillet, 4F Position	**177**
Procedure 80	V-Groove Butt Joint, 4G Position	**179**

Gas Tungsten Arc Welding of Aluminum — **181**

Procedure 81	Running a Flat Pad of Stringers	**183**
Procedure 82	T-Joint Fillet, 1F Position	**185**
Procedure 83	Square-Groove Butt Joint, 1G Position	**187**
Procedure 84	T-Joint Fillet, 2F Position	**189**
Procedure 85	Square-Groove Butt Joint, 2G Position	**191**
Procedure 86	T-Joint Fillet, 3F Position	**193**
Procedure 87	Square-Groove Butt Joint, 3G Position	**195**
Procedure 88	T-Joint Fillet, 4F Position	**197**
Procedure 89	Square-Root Butt Joint, 4G Position	**199**

GAS METAL ARC WELDING PROCESS — **201**

Short-Circuiting Arc Welding of Steel — **203**

Procedure 90	Running a Flat Pad of Stringer Beads	**205**
Procedure 91	T-Joint Fillet, 1F Position	**207**
Procedure 92	V-Groove Butt Joint, 1G Position	**209**
Procedure 93	T-Joint Fillets, 2F Position	**211**
Procedure 94	V-Groove Butt Joint, 2G Position	**213**
Procedure 95	T-Joint Fillet, 3F Position, Upward Welding	**215**
Procedure 96	T-Joint Fillet, 3F Position, Downward Welding	**217**
Procedure 97	V-Groove Butt Joint, 3G Position with Backup Bar, Upward Welding	**219**
Procedure 98	V-Groove Butt Joint, 3G Position, Downward Welding	**221**

Procedure 99	T-Joint Fillet, 4F Position	**223**
Procedure 100	V-Groove Butt Joint, 4G Position	**225**

Spray Transfer Welding of Steel — 227

Procedure 101	T-Joint Fillet, 1F Position	**229**
Procedure 102	V-Groove Butt Joint, 1G Position with Backup Bar	**231**
Procedure 103	T-Joint Fillet, 2F Position	**233**
Procedure 104	V-Groove Butt Joint, 2G Position with Backup Bar	**235**

FLUX-CORED ARC WELDING PROCESS — 237

Flux-Cored Arc Welding of Steel — 239

Procedure 105	Running Flat Pad of Stringers	**241**
Procedure 106	T-Joint Fillet, 1F Position	**243**
Procedure 107	V-Groove Butt Joint, 1G Position with Backup Bar	**245**
Procedure 108	T-Joint Fillet, 2F Position	**247**
Procedure 109	V-Groove Butt Joint, 2G Position with Backup Bar	**249**
Procedure 110	T-Joint Fillet, 3F Position	**251**
Procedure 111	V-Groove Butt Joint, 3G Position with Backup Bar	**253**
Procedure 112	T-Joint Fillet, 4F Position	**255**
Procedure 113	V-Groove Butt Joint, 4G Position with Backup Bar	**257**

OXYFUEL GAS WELDING AND BRAZING

Oxyacetylene Welding (OAW) of Carbon Steel

The Procedures in this section of the workbook will aid you in developing your skills in oxyfuel gas welding and brazing. You will use these skills as you advance to other welding processes.

The relatively low heat needed to perform oxyfuel gas welding procedures permits you to observe the weld puddle at close distance. Thus, you will learn how to better manipulate the weld puddle.

Braze welding is a welding process used routinely in the repair of broken machinery. The Procedures presented in the latter part of this section will aid you in developing these skills. It is important that you complete the Procedures in the order that they are presented here. The sequence of the Procedures in the workbook will ensure that you develop the necessary skills to master more and more complex welding procedures.

GENERAL INSTRUCTIONS

For each of the procedures in this section, follow the steps listed below:

1. Assemble the oxyacetylene welding outfit according to the manufacturer's instructions or the instructions of your teacher.
2. Select the correct tip size from the manufacturer's chart and adjust the regulators to the correct working pressure.
3. Check all connections with a leak detection fluid or soapy water to ensure tightness.

Safety Note: Do not neglect this important step. Oxyfuel mixtures are highly explosive and extremely flammable. Failure to check for leaks could result in serious bodily injury and severe property damage.

4. Clean the surface of the plate you are going to weld on with a wire brush to remove mill scale and oxides, then place it on the welding table.
5. Check the area around you and your welding table to make sure it is free of all combustible materials.
6. Position yourself comfortably in front of your welding station. A standing position, feet slightly apart, is best. This position allows maximum body movement to aid in controlling the torch.

MATERIALS AND EQUIPMENT

For this section the material and equipment will be as follows:

1. Proper clothing
2. Safety glasses and welding goggles with a number 4 or 5 lens
3. Leather gloves
4. Oxyacetylene torch, tips, and regulators
5. Pliers or tongs
6. Spark lighter
7. Wire brush

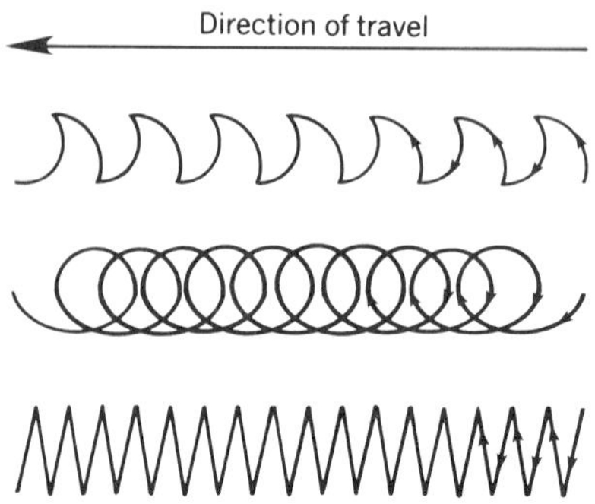

4 OXYFUEL GAS WELDING AND BRAZING

FIGURE 1-1

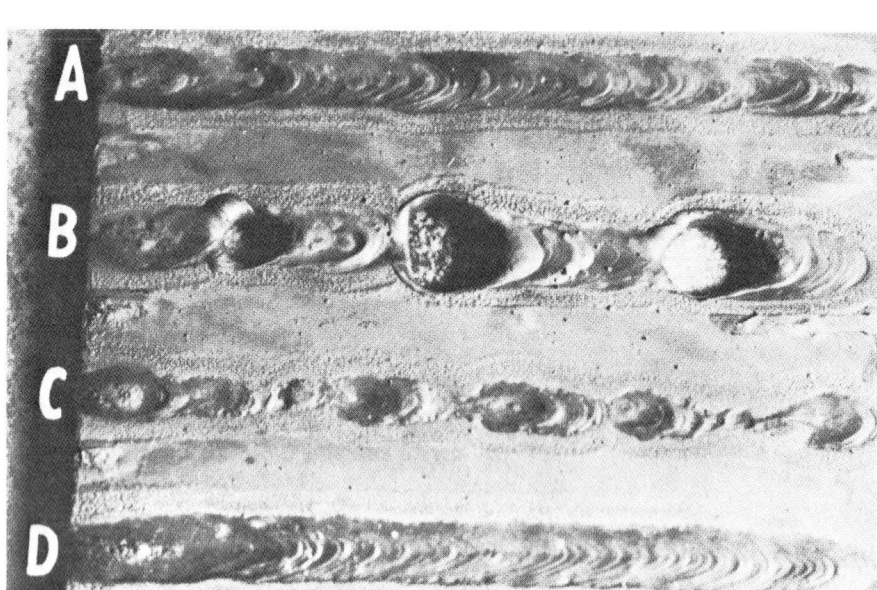

FIGURE 1-2
(From *Welding Fundamentals and Procedures* by Jerry Galyen, Garry Sear, and Charles A. Tuttle. Copyright © 1984 by John Wiley & Sons, Inc. Reprinted by permission of John Wiley & Sons, Inc.)

NAME	Beads without filler metal	POSITION	Flat
ELECTRODE None	DIAMETER None	FLAME TYPE	Neutral
MATERIAL	1 pc. 3/32" × 3" × 6" carbon steel		
PASSES Multiple	BEAD String	TIME (SEE INSTRUCTOR)	

Procedure 1

Flat Beads Without Filler Metal

OBJECTIVE

Upon completion of this procedure you should be able to weld beads in the flat position without filler metal.

Text Reference: Section II, Lesson 5A; page 154.

PROCEDURE

1. Light the torch and adjust it to obtain a neutral flame. Be sure your safety glasses and goggles are in place.
2. Lower the torch until the inner core of the flame is about 3/32 in. above the surface on the plate. The torch tip should be at a 40 to 50 degree angle pointing toward the direction of travel. (Pointing toward the left for a right-handed person and toward the right for a left-handed person.)
3. Hold the torch there until a round puddle develops. Use one of the side-to-side motions shown in Figure 1-1 to establish the puddle.
4. When the puddle has been established, progress at a uniform rate of travel toward the left (right for left-handed welders) being careful to adjust the rate of travel to obtain a bead of uniform width.
5. Continue to run beads until the plate is filled up.
6. When the plate has cooled examine the bead appearance. (See Figure 1-2.)

Safety Note: Do not handle the metal with bare hands. It may be hot or have sharp edges that will burn or cut your hand. Use your pliers or tongs to handle the metal.

6 OXYFUEL GAS WELDING AND BRAZING

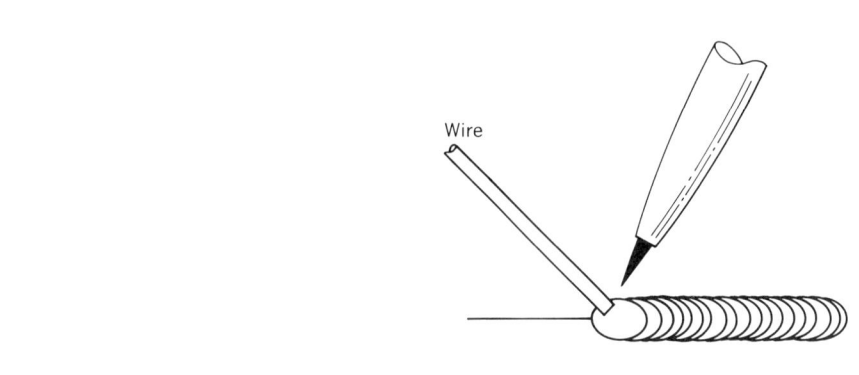

FIGURE 2-1

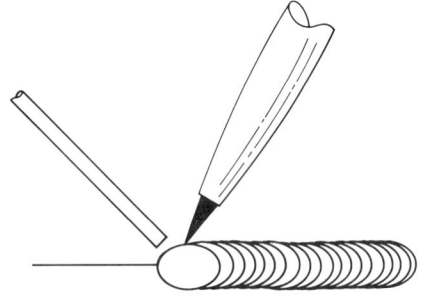

FIGURE 2-2

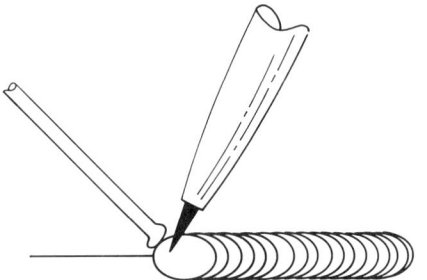

FIGURE 2-3

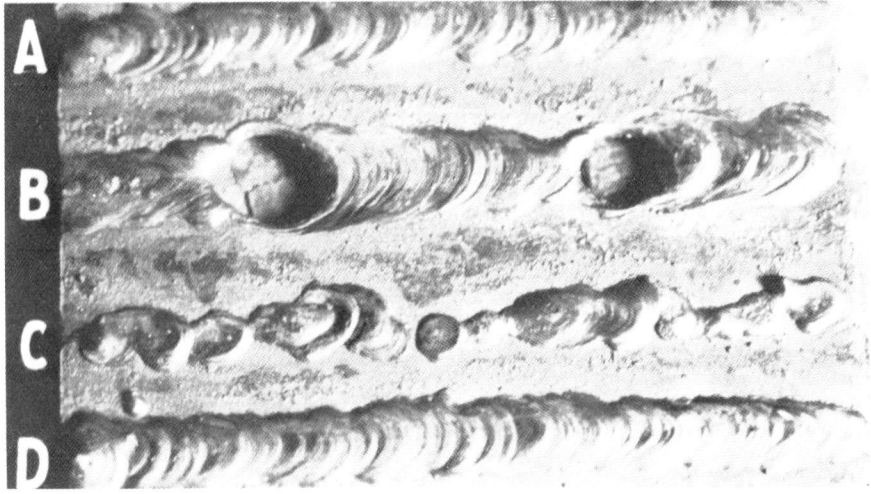

FIGURE 2-4
(From *Welding Fundamentals and Procedures* by Jerry Galyen, Garry Sear, and Charles A. Tuttle. Copyright © 1984 by John Wiley & Sons, Inc. Reprinted by permission of John Wiley & Sons, Inc.)

NAME	Beads with filler metal		POSITION	Flat
ELECTRODE Carbon steel	DIAMETER	3/32″	FLAME TYPE	Neutral
MATERIAL	1 pc. 3/32″ × 3″ × 6″ carbon steel			
PASSES Multiple	BEAD	String	TIME (SEE INSTRUCTOR)	

Procedure 2

Flat Beads with Filler Metal

OBJECTIVE

Upon completion of this procedure you should be able to weld beads in the flat position using filler metal.

Text Reference: Section II, Lesson 5A; page 156.

PROCEDURE

1. Light the torch and adjust it to obtain a neutral flame. Be sure your safety glasses and goggles are in place.
2. Lower the torch to the plate and establish a weld puddle. Hold the filler metal in your left hand. As the puddle develops, move the end of the filler metal to the leading edge of the puddle and allow the wire to melt off and mix with the puddle. (See Figure 2-1.) Withdraw the wire slightly, keeping the end of the wire in the secondary flame envelope to protect it from the atmosphere. (See Figure 2-2.) Do not keep it too close to the flame or the end will melt and ball up. This large ball makes it very difficult to add filler metal to the puddle. (See Figure 2-3.)
3. Continue to add filler metal to the leading edge of the puddle by dabbing it in the puddle as the torch is advanced along the sheet. Coordinate the rate of travel and the addition of filler metal, paying attention to maintain a uniform bead width and height.
4. When nearing the edge of the sheet gradually withdraw the flame and add filler metal to fill the crater.
5. Continue to run beads until the plate is filled up.
6. When the plate has cooled, examine the bead appearance. (See Figure 2-4.)

Note: Filler metals are classified by the American Welding Society* in filler metal specifications. Filler metal rods for gas welding range in diameter from $1/16$ to $3/16$ in. and are usually 36 in. in length. Mild steel welding rods are often copper coated to inhibit rusting. The size of the rod you use will depend on the thickness of the metal to be welded and the position it is to be welded in. Too large a rod will cool the puddle, decreasing penetration and fusion. A rod that is too small will cause excessive heat, resulting in too much penetration and too wide a weld bead.

*American Welding Society, 550 N.W. LeJeune Road, Miami, FL 33135

8 OXYFUEL GAS WELDING AND BRAZING

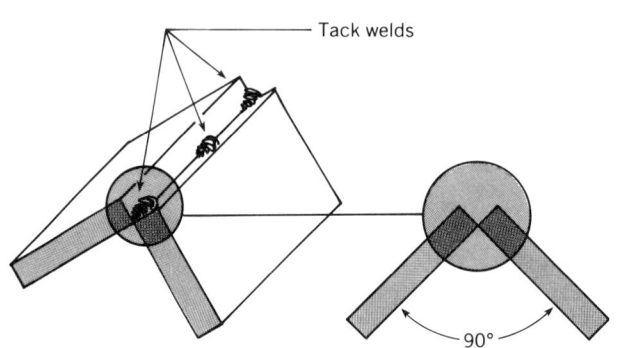

FIGURE 3-1

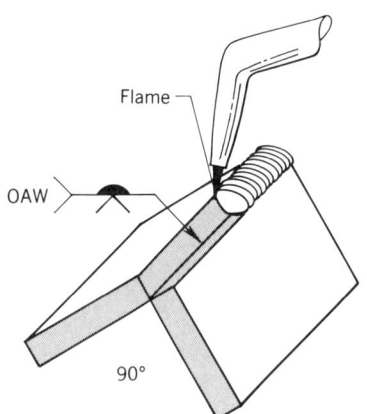 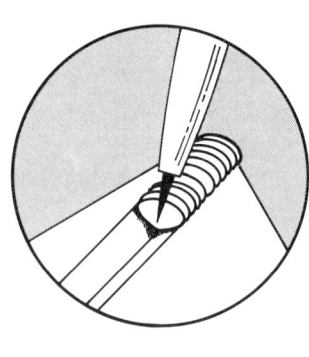

FIGURE 3-2

NAME			POSITION	
	Corner weld without filler			Flat
ELECTRODE		DIAMETER	FLAME TYPE	
	None	None		Neutral
MATERIAL				
	2 pcs. 3/32" × 1 1/2" × 6" carbon steel			
PASSES		BEAD	TIME (SEE INSTRUCTOR)	
	Single	Weave		

Procedure 3

Flat Outside-Corner Joint

OBJECTIVE

Upon completion of this lesson you should be able to weld a flat corner joint without filler metal.

Text Reference: Section II, Lesson 5A; page 157.

PROCEDURE

1. Light the torch and adjust it to obtain a neutral flame. Be sure your safety glasses and goggles are in place.
2. Tack the two pieces together as shown in Figure 3-1.
3. Lower the torch to the joint so the inner cone is about $1/16$ in. above the surface of the joint. Move in a slight circular motion until you establish a puddle. Observe the puddle and make sure you melt the groove faces and the root of the joint.
4. Continue using one of the weaving motions you have learned. Make sure that the puddle forms behind the flame to complete the weld. (See Figure 3-2.) Regulate your speed of travel so the bead has an even width. Too rapid a travel speed will not allow the bead to form. Too slow a travel speed will cause an excessively large puddle that will fall through, causing a large hole.
5. Continue along the joint until you reach the end. When you reach the end continue your weaving motion and gradually lift the flame away from the surface until the puddle solidifies.
6. After the weld cools, clean it with your wire brush and examine it.

Observe, as you practice, that the tip to work distance can be used as an aid to control the weld puddle. If the puddle becomes too large, increase the flame to puddle distance to cool off the puddle. If you get too close the sparks and spatter that come off of the weld puddle can enter the tip and clog it, causing backfires or backflashes. Also be sure to set the correct gas pressures. Excessive pressure will cause a strong flame that will make weld puddle difficult to control.

10 OXYFUEL GAS WELDING AND BRAZING

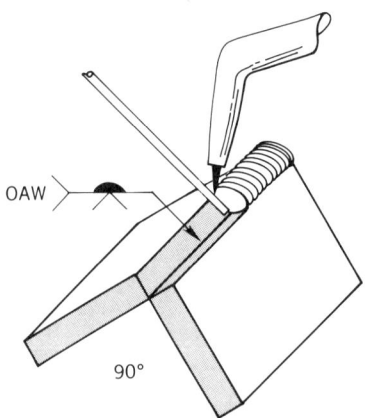

FIGURE 4-1

NAME			POSITION	
	Corner weld with filler			Flat
ELECTRODE		DIAMETER	FLAME TYPE	
Carbon steel		3/32"		Neutral
MATERIAL				
2 pcs. 3/32" × 1 1/2" × 6" carbon steel				
PASSES		BEAD	TIME (SEE INSTRUCTOR)	
Single		Weave		

Procedure 4

Flat Outside-Corner Joint with Filler Metal

OBJECTIVE

Upon completion of this lesson you should be able to weld flat corner joints using filler metal.

Text Reference: Section II, Lesson 5A; page 158.

PROCEDURE

1. Light the torch and adjust it to obtain a neutral flame. Be sure your safety glasses and goggles are in place.
2. Tack the two pieces together as shown in Figure 4-1.
3. Lower the torch to the joint and establish a puddle. Carefully note when you establish the puddle that you melt all areas of the joint, including the root. This will determine your penetration. Be careful not to allow the torch to remain in one spot too long or you will develop a puddle so large it will fall through. The resulting hole is extremely difficult to repair. You must watch the puddle carefully to make sure the surfaces of the groove and the root are welded, otherwise you will not achieve full penetration.
4. Progress along the joint adding filler metal as you learned earlier, make sure that you obtain complete penetration and watch that your bead width and reinforcement are uniform.
5. Weld the complete joint, be careful to withdraw the flame slowly at the end of the joint so you do not melt the end. While withdrawing the flame, add filler rod to the puddle to fill the crater.

12 OXYFUEL GAS WELDING AND BRAZING

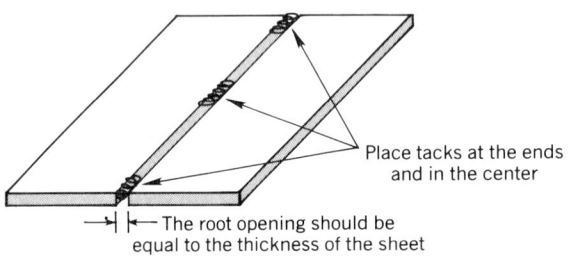

FIGURE 5-1

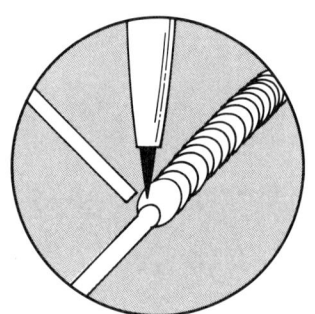
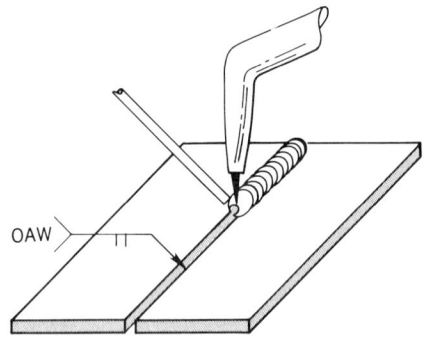

FIGURE 5-2

NAME	Flat butt with filler	POSITION	Flat
ELECTRODE Carbon steel	DIAMETER 3/32"	FLAME TYPE	Neutral
MATERIAL	2 pcs. 3/32" × 3" × 6" carbon steel		
PASSES Single	BEAD Weave	TIME (SEE INSTRUCTOR)	

Procedure 5

Flat Square-Groove Butt Joint with Filler Metal

OBJECTIVES

Upon completion of this lesson you should be able to weld flat butt joints.

Text Reference: Section II, Lesson 5A; page 159.

PROCEDURE

1. Tack the two pieces together as shown in Figure 5-1.

2. Lower the torch to the joint and establish a puddle. To ensure penetration use the keyhole technique. The keyhole technique involves allowing the puddle to penetrate through the joint until a keyhole (see Figure 5-2) is established.

3. Once you have established a keyhole begin adding filler metal to the leading edge of the puddle, be careful to control rate of travel along the joint and rate of filler metal addition to produce an even, consistent bead.

14 OXYFUEL GAS WELDING AND BRAZING

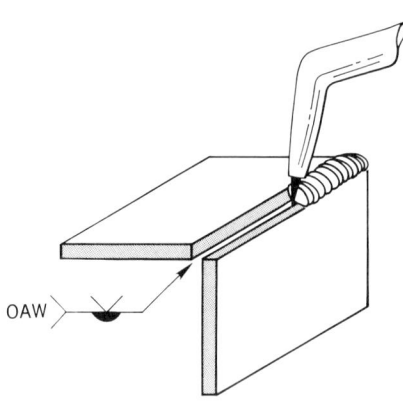

FIGURE 6-1

NAME			POSITION	
	Open corner			**Horizontal**
ELECTRODE		DIAMETER	FLAME TYPE	
Carbon steel		³⁄₃₂"		**Neutral**
MATERIAL				
	2 pcs. ³⁄₃₂" × 1 ½" × 6" carbon steel			
PASSES		BEAD	TIME (SEE INSTRUCTOR)	
Single		**Weave**		

Procedure 6

Horizontal Outside-Corner Joint

OBJECTIVE

Upon completion of this lesson you should be able to weld open corner joints in the horizontal position.

Text Reference: Section II, Lesson 5B; page 159.

PROCEDURE

1. Light the torch and adjust it to obtain a neutral flame. Be sure your safety glasses and goggles are in place.
2. Tack the two pieces together and position as shown in Figure 6-1.
3. Lower the torch to the joint so the tip of the inner core is about 1/16 in. above the surface of the joint. Move the flame in one of the weave patterns you have learned until you establish a puddle.
4. Continue the weld using the weave pattern. Regulate your travel speed to produce a bead of consistent width with even ripples.
5. When you reach the end of the joint gradually lift the flame away from the surface until the puddle solidifies.
6. Set up another horizontal corner weld as shown in Figure 6-1. Begin the weld as in Step 2, but this time add the filler metal to the puddle as you progress along the joint. Make sure you melt the filler metal completely and mix it with the weld puddle thoroughly before advancing the flame along the joint.
7. Allow the plate to cool. Clean it, then examine it for joint penetration and face appearance.

Safety Note: If you force cool your welds by immersing them in water be sure to wear your gloves and stand off to the side to prevent steam burns to your hands or face.

16 OXYFUEL GAS WELDING AND BRAZING

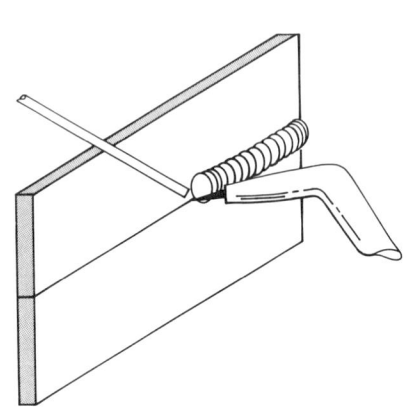

FIGURE 7-1

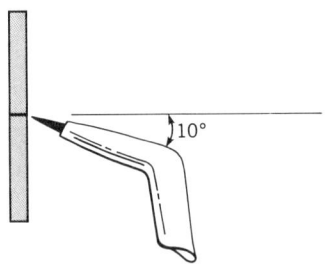

FIGURE 7-2

NAME	Square groove butt	POSITION	Horizontal
ELECTRODE Carbon steel	DIAMETER 3/32"	FLAME TYPE	Neutral
MATERIAL	2 pcs. 3/32" × 3" × 6" carbon steel		
PASSES Single	BEAD String	TIME (SEE INSTRUCTOR)	

Procedure 7

Horizontal Square-Groove Butt Joint

OBJECTIVE

Upon completion of this lesson you should be able to weld square-groove butt welds in the horizontal position.

Text Reference: Section II, Lesson 5B; page 160.

PROCEDURE

1. Tack the two pieces together as shown in Figure 7-1.
2. Position the plate so that the weld joint is in the horizontal 2G position.
3. Position the torch so that the tip is pointing up at about a 10-degree angle. (See Figure 7-2.) The flame angle will help to control the weld puddle to provide even reinforcement and to prevent undercut at the toe of the weld on the top side of the joint.
4. If the puddle becomes too large and begins to sag, you can do two things.
   a. You can add less filler metal and work the puddle you have with the flame.
   b. You can move the flame farther away from the plate. This will reduce the amount of heat put into the joint.
5. Complete the joint being careful to fill the crater as you complete the weld.

18 OXYFUEL GAS WELDING AND BRAZING

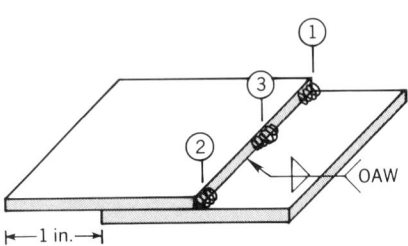

FIGURE 8-1

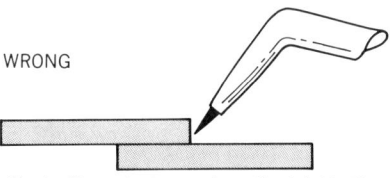

WRONG

Heat will conduct away from the joint in the bottom sheet, causing the top to heat more quickly

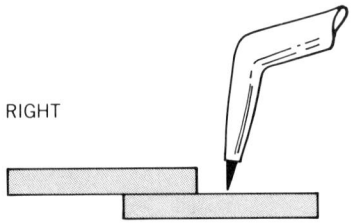

RIGHT

Directing the flame toward the bottom sheet will make the melting more even

FIGURE 8-2

NAME	Fillet weld	POSITION	Horizontal
ELECTRODE Carbon steel	DIAMETER 3/32"	FLAME TYPE	Neutral
MATERIAL	2 pcs. 3/32" × 3" × 6" carbon steel		
PASSES Single	BEAD String	TIME (SEE INSTRUCTOR)	

Procedure 8

Horizontal Fillet Joint Position

OBJECTIVE

Upon completion of this lesson you should be able to weld fillet welds in the horizontal position.

Text Reference: Section II, Lesson 5B; page 160.

PROCEDURE

1. Tack the two pieces together as shown in Figure 8-1. Tack both sides.
2. Lower the torch to the joint and establish a puddle. As shown in Figure 8-2, you must angle the torch differently to obtain equal melting at the joint.
3. Add filler rod as you progress along the joint. Check that both legs and the root are being fused, the bead width is even, and the face is flat or slightly convex. Weld both sides.

20 OXYFUEL GAS WELDING AND BRAZING

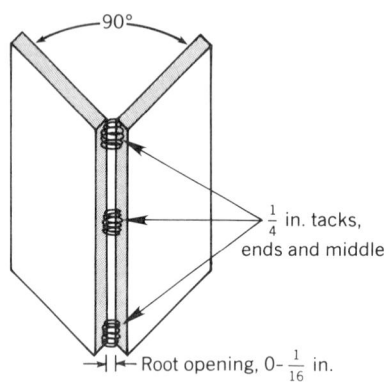

FIGURE 9-1

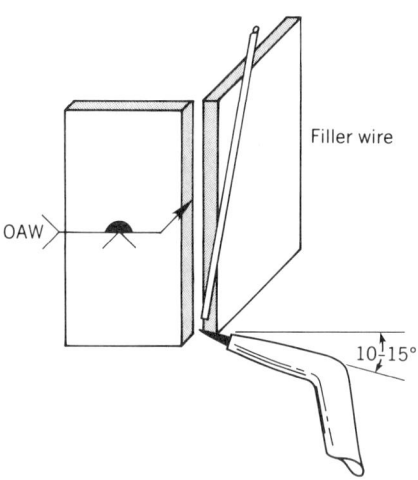

FIGURE 9-2

NAME	Open corner	POSITION	Vertical up
ELECTRODE Carbon steel	DIAMETER 3/32"	FLAME TYPE	Neutral
MATERIAL	2 pcs. 3/32" × 1 1/2" × 6" carbon steel		
PASSES Single	BEAD Weave	TIME (SEE INSTRUCTOR)	

Procedure 9

Vertical Outside-Corner Joint, Uphill Weld

OBJECTIVE

Upon completion of this lesson you should be able to weld an open-corner joint vertically up.

Text Reference: Section II, Lesson 5C; page 161.

PROCEDURE

1. Light the torch and adjust it to obtain a neutral flame. Be sure your safety glasses and goggles are in place.
2. Tack the two pieces together and position as shown in Figure 9-1.
3. Move the tip to the joint and position the torch as shown in Figure 9-2.
4. Establish a puddle, observing that the side walls are molten to the root of the joint.
5. When a puddle is established, add the filler wire to the leading edge of the puddle. (See Figure 9-2.) Add the wire as needed. Do not use a continuous feed.
6. Use a side-to-side weave and progress vertically up, taking care to keep the bead width uniform and the contour even.
7. Progress to the top of the joint, then gradually pull away the torch, adding filler wire to build up the crater.
8. When the plate has cooled clean it, examine it for joint penetration and face appearance.

22 OXYFUEL GAS WELDING AND BRAZING

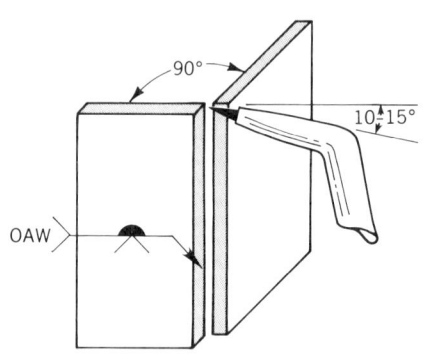

FIGURE 10-1

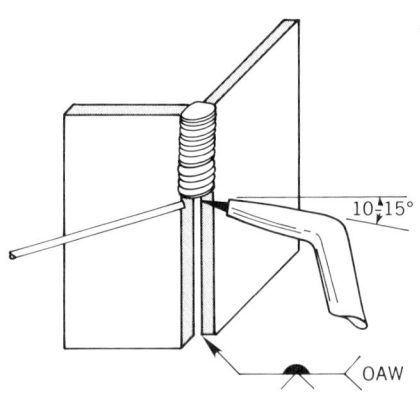

FIGURE 10-2

NAME	Open corner	POSITION	Vertical down
ELECTRODE Carbon steel	DIAMETER 3/32"	FLAME TYPE	Neutral
MATERIAL	2 pcs. 3/32" × 1 1/2" × 6" carbon steel		
PASSES Single	BEAD Weave	TIME (SEE INSTRUCTOR)	

Procedure 10

Vertical Outside-Corner Joint, Downhill Weld

OBJECTIVE

Upon completion of this lesson you should be able to weld an open corner joint in the vertical down position.

Text Reference: Section II, Lesson 5C; page 162.

PROCEDURE

1. Tack the two pieces together and position as shown in Figure 10-1.
2. Move the tip to the joint and position the torch as shown in Figure 10-1.
3. Establish a puddle, be careful to make sure the side walls are melted completely to the root of the joint.
4. Begin moving downward at a rate of speed that will ensure a complete weld.

Caution: A travel speed that is too slow will result in a large puddle that will drop out leaving a hole. A travel speed that is too fast will result in only partial melting of the joint. If this occurs, the metal will not flow behind the flame to form a completed weld.

5. Use a side-to-side motion to help the weld form. If you need more control over the weld puddle, increase the torch angle in order to use the flame to help support the puddle.
6. Complete the downhill progression using a weaving motion until the joint is complete.
7. Tack two more pieces together as shown in Figure 10-1.
8. Establish a puddle as you did in Step 3, but this time as you begin the downhill progression begin adding filler metal as shown in Figure 10-2.
9. Complete the weld being careful to watch your filler wire addition and travel speed to provide an even, consistent bead.
10. Upon completion of the welds clean them and examine them for root penetration and face appearance.

24 OXYFUEL GAS WELDING AND BRAZING

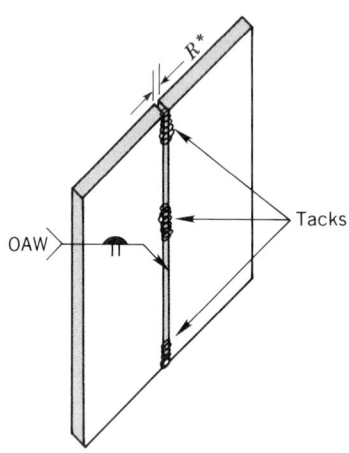

*Root opening should be up to half the thickness of the sheet used

FIGURE 11-1

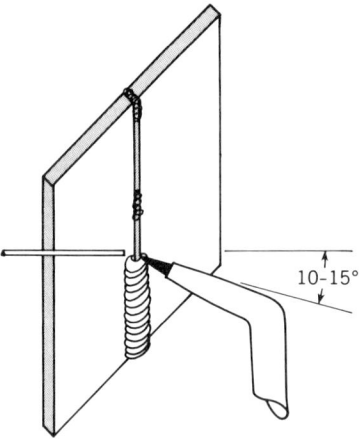

FIGURE 11-2

NAME Square-groove butt	POSITION Vertical up
ELECTRODE Carbon steel DIAMETER 3/32"	FLAME TYPE Neutral
MATERIAL 2 pcs. 3/32" × 3" × 6" carbon steel	
PASSES Single BEAD Weave	TIME (SEE INSTRUCTOR)

Procedure 11

Vertical Square-Groove Butt Joint

OBJECTIVE

Upon completion of this lesson you should be able to weld square-groove butt joints in the vertical position.

Text Reference: Section II, Lesson 5C; page 162.

PROCEDURE

1. Tack two pieces together and position as shown in Figure 11-1.
2. Move the torch to the bottom of the groove and establish a weld puddle. Continue to develop the puddle until you see the keyhole that indicates complete penetration has been obtained.
3. When you achieve the desired penetration begin adding filler metal and progress upward. (See Figure 11-2.) Use one of the weaving motions you have learned to make sure the puddle fuses to both sides of the joint.
4. Progress upward at an even rate of travel, while adding filler metal to produce a bead of even width with good contour and appearance.
5. End the weld being sure to taper off at the end to fill the crater and complete the joint.

Caution: Craters are often the source of cracks and blisters due to inadequate fill.

26 OXYFUEL GAS WELDING AND BRAZING

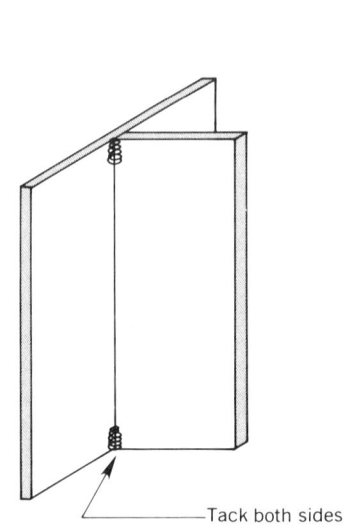

FIGURE 12-1

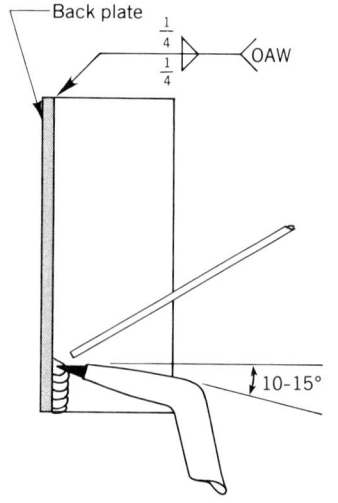
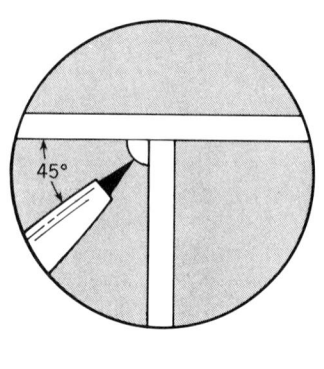

FIGURE 12-2

NAME			POSITION	
	Fillet			Vertical up
ELECTRODE		DIAMETER	FLAME TYPE	
Carbon steel		3/32"		Neutral
MATERIAL				
	1 pc. 3/32" × 3" × 6" carbon steel			
	1 pc. 3/32" × 1 1/2" × 6" carbon steel			
PASSES		BEAD	TIME (SEE INSTRUCTOR)	
Single		Bead Weave		

Procedure 12

Vertical Fillet Joint

OBJECTIVE

Upon completion of this lesson you should be able to weld vertical fillet welds.

Text Reference: Section II, Lesson 5C; page 163.

PROCEDURE

1. Tack the assembly together and position it as shown in Figure 12-1.
2. Position the torch as shown in Figure 12-2 and establish a puddle.

Note: Extreme care must be taken so that melting occurs all the way into the root of the joint. The side walls will melt first, but do not begin adding filler metal until the root is molten.

3. Begin the vertical progression, adding filler metal as shown in Figure 12-2, keeping the rate of travel and filler metal addition even so an even bead is deposited.
4. Apply a ¼ in. leg fillet on both sides of the joint as indicated by the welding symbol.
5. When you have completed the welds, clean them and examine them.

28 OXYFUEL GAS WELDING AND BRAZING

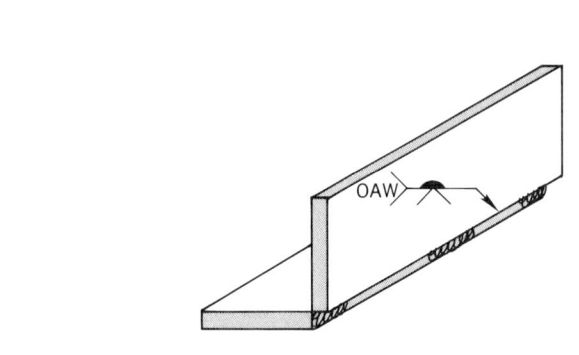

FIGURE 13-1

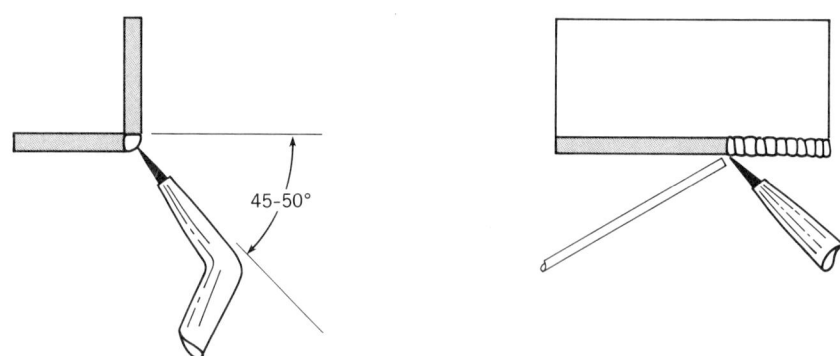

FIGURE 13-2

NAME	Corner weld		POSITION	Overhead
ELECTRODE Carbon steel	DIAMETER	3/32"	FLAME TYPE	Neutral
MATERIAL	2 pcs. 3/32" × 1 1/2" × 6" carbon steel			
PASSES Single	BEAD	Weave	TIME (SEE INSTRUCTOR)	

Procedure 13

Overhead Outside-Corner Joint

OBJECTIVE

Upon completion of this lesson you should be able to weld overhead outside-corner joints.

Text Reference: Section II, Lesson 5D; page 164.

PROCEDURE

1. Tack the two pieces together and position as shown in Figure 13-1.
2. Bring the torch to the joint and establish a puddle. Look for the keyhole indicating complete melting.
3. Use the weave technique most comfortable for you and begin progressing along the joint evenly.
4. Fuse the entire joint looking for complete penetration and even bead width.
5. Tack two more pieces together as shown in Figure 13-1 and again establish a puddle.
6. When a puddle has been established begin adding filler metal as shown in Figure 13-2.
7. When you have completed the beads clean them and examine them.

30 OXYFUEL GAS WELDING AND BRAZING

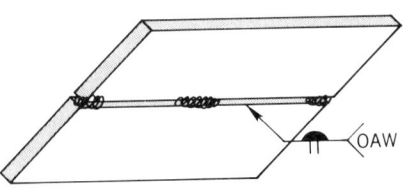

FIGURE 14-1

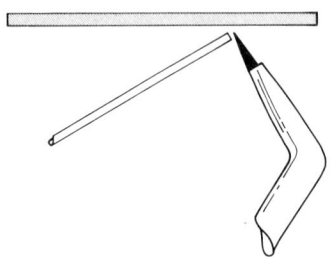

FIGURE 14-2

NAME		POSITION	
	Square-groove weld		Overhead
ELECTRODE	DIAMETER	FLAME TYPE	
Carbon steel	3/32″		Neutral
MATERIAL	2 pcs. 3/32″ × 3″ × 6″ carbon steel		
PASSES	BEAD	TIME (SEE INSTRUCTOR)	
Single	Weave		

Procedure 14

Overhead Square-Groove Butt Joint

OBJECTIVE

Upon completion of this lesson you should be able to weld overhead square-groove butt joints.

Text Reference: Section II, Lesson 5D; page 164.

PROCEDURE

1. Tack the two pieces together and position as shown in Figure 14-1.
2. Bring the torch to the plate and establish a puddle; be careful to obtain the keyhole. Position the torch as shown in Figure 14-2.
3. Add filler metal as shown in Figure 14-2. Pay attention to bead width and surface contour.

Note: The flame can be used to aid in controlling the puddle.

4. Complete the joint and clean and examine it.

32 OXYFUEL GAS WELDING AND BRAZING

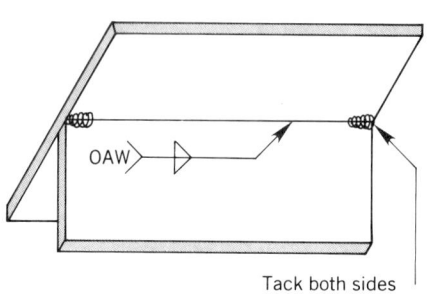

FIGURE 15-1

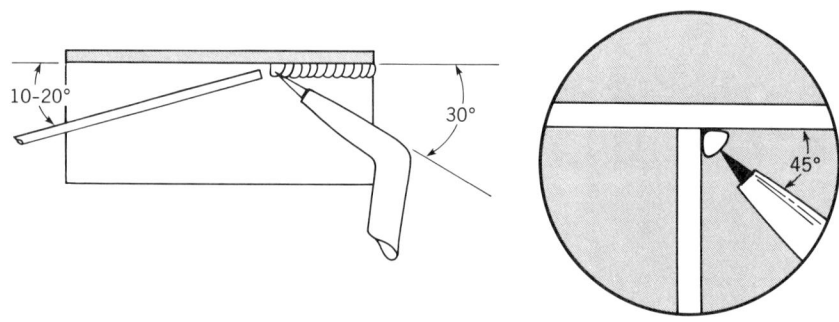

FIGURE 15-2

NAME Fillet weld	POSITION Overhead
ELECTRODE Carbon steel　　　　DIAMETER　　　　3/32"	FLAME TYPE Neutral
MATERIAL 1 pc. 3/32" × 3" × 6" carbon steel 1 pc. 3/32" × 1 1/2" × 6" carbon steel	
PASSES Single　　　　BEAD 　　　　Weave	TIME (SEE INSTRUCTOR)

Procedure 15

Overhead Fillet Joint

OBJECTIVE

Upon completion of this lesson you should be able to weld overhead fillets.

Text Reference: Section II, Lesson 5D; page 165.

PROCEDURE

1. Tack the two pieces together and position as shown in Figure 15-1.
2. Bring the torch to the joint and establish a puddle; be careful to develop the molten puddle to the root of the joint.

Note: By now you have observed that in T-joints, the side walls become molten much quicker than the root. The natural tendency is to begin adding filler metal to avoid excessive side wall melting. Doing so usually results in incomplete root penetration, since the puddle will bridge across between the two members of the joint, and not reach the root. Do not begin adding the filler metal until the root of the joint has begun to melt.

3. Continue along the joint using the torch and filler metal, positions shown in Figure 15-2. Use one of the weave-bead techniques you have learned in the earlier lessons.
4. Weld the other side of the joint. When you have completed the welds, clean and examine them.

Oxyacetylene Braze-Welding

Braze-welding may be defined as a process by which metals are joined using a filler metal that melts above 840 degrees F. but below the melting point of the base metal. Unlike brazing the filler metal is not drawn into the joint by capillary action. The filler metal is deposited in layers. The filler metal wets the surface of the base metal, but does not mix with it as in fusion welding. The surfaces to be joined must be clean and a flux must be used. The flux is applied usually by one of two methods. In the first method the end of the brazing rod is heated slightly in the flame, then inserted into a container of dry granulated flux. Some of the flux will adhere to the warmed filler rod, and will provide the necessary fluxing action for the braze-weld. In the second method, the filler metal may be purchased with a flux coating. In this case flux addition to the puddle is continuous.

GENERAL INSTRUCTIONS

The following steps are necessary to complete the procedures in this section:

1. Grind the surface of the plate to remove mill scale. The surfaces to be joined by braze-welding must be clean or a good bond will not be achieved.
2. Assemble the oxyacetylene welding outfit according to the manufacturer's instructions or the instructions of your teacher.
3. Select the correct tip size from the manufacturer's chart and adjust the regulators to the correct working pressure.
4. Check all connections with a leak detection fluid or soapy water to ensure tightness.
5. Clean the surface of the plate you are going to weld on with a wire brush to remove mill scale and oxides, and then place it on the welding table.
6. Check the area around you and your welding table to make sure it is free of all combustible materials.
7. Position yourself comfortably in front of your welding station. A standing position, feet slightly apart, is best. This position allows maximum body movement to aid in controlling the torch.

MATERIALS AND EQUIPMENT

For this section the material and equipment will be as follows:

1. Proper clothing
2. Safety glasses and welding goggles with a number 4 or 5 lens
3. Leather gloves
4. Oxyacetylene torch, tips, and regulators
5. Pliers or tongs
6. Spark lighter
7. Wire brush
8. Brazing rod with $3/32$ in. diameter (flux coated or with a can of flux)

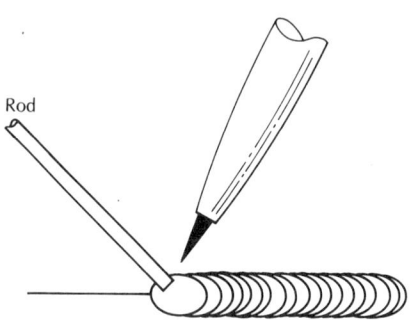

FIGURE 16-1

NAME	Running beads	POSITION	Flat
ELECTRODE Brazing rod	DIAMETER 3/32"	FLAME TYPE	Neutral or slightly oxidizing
MATERIAL	1 pc. ¼" × 6" × 6" carbon steel		
PASSES Single	BEAD Weave	TIME (SEE INSTRUCTOR)	

Procedure 16

Running Flat Beads

OBJECTIVE

Upon completion of this lesson you should be able to braze-weld beads in the flat position.

Text Reference: Section II, Lesson 6A; page 167.

PROCEDURE

1. Light the torch and adjust it to obtain a reducing flame. Be sure your safety glasses and goggles are in place. In braze-welding you should avoid any chance of contamination in the area you intend to join, since this will make it very difficult for the braze metal to weld the metal properly. For this reason it is best to adjust the flame to obtain a neutral or slightly oxidizing flame.
2. Lower the torch so the inner cone is just above the surface of the plate.
3. Use a circular motion to heat the plate to a dull red.
4. Bring the brazing rod to the surface and melt off a bit of the rod with the flame. (See Figure 16-1.)
5. Use the flame to manipulate the braze metal. If the plate is at the proper temperature the metal will immediately wet the surface of the plate.

Note: The term "wet" or "wetting" can be defined as the ability of the filler metal, using the proper flux, to coat the surface you desire to join without causing that surface to become molten.

6. Manipulate the flame to work the filler metal so it wets the surface and gains the desired buildup and bead width. Heat control is important. Insufficient heat will not allow the braze metal to wet the surface or flow properly. Excessive heat will cause the braze metal to flow excessively and not allow it to build up. You can control the amount of the heat you apply by varying the distance between the torch and the plate.
7. Continue the bead to the end of the plate. Because the plate will become very hot, it is a good idea to quench the plate after every bead. This will also help remove the flux from the bead.
8. Clean the bead and examine it. Be sure to wear your safety glasses when you remove the flux. Chip the flux and wire brush it thoroughly.
9. Fill the plate with beads and examine them.

38 OXYFUEL GAS WELDING AND BRAZING

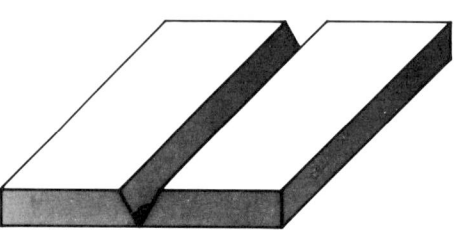

FIGURE 17-1

FIGURE 17-2

NAME	V-groove butt	POSITION	Flat
ELECTRODE Brazing rod	DIAMETER 3/32"	FLAME TYPE	Neutral or slightly oxidizing
MATERIAL 2 pcs. ¼" × 3" × 6" carbon steel, with a 30-degree bevel on one 6 in. side			
PASSES Single	BEAD Weave	TIME (SEE INSTRUCTOR)	

Procedure 17

Flat V-Groove Butt Joint

OBJECTIVE

Upon completion of this lesson you should be able to braze-weld flat V-groove butt joints.

Text Reference: Section II, Lesson 6A; page 167.

PROCEDURE

1. Grind the surfaces to be joined.
2. Tack the assembly together as shown in Figure 17-1.
3. Heat the metal to a dull cherry color. When the proper temperature has been reached, begin adding braze metal. Observe that the braze metal wets through to the root of the joint. When this occurs begin the first pass being sure to wet all the joint surfaces.
4. Complete the first pass, be careful to obtain a concave weld face.
5. Complete the joint, adding passes two and three as shown in Figure 17-2.
6. Clean the plate thoroughly and examine it. If your weld is done properly you will observe filler metal that has penetrated through the root of the joint and has wet all the surfaces. The face of the weld should be smooth and have even ripples.

40 OXYFUEL GAS WELDING AND BRAZING

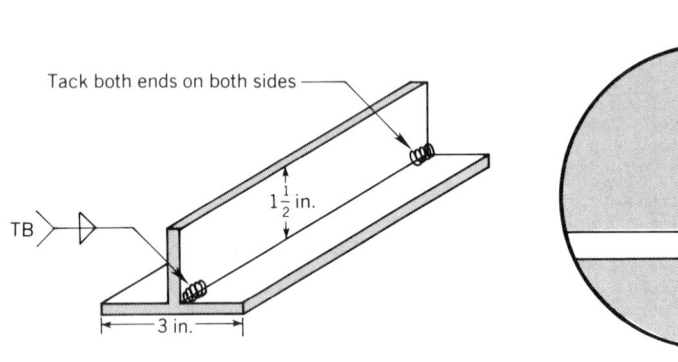

FIGURE 18-1

NAME	T-fillet	POSITION	Horizontal
ELECTRODE Brazing rod	DIAMETER 3/32"	FLAME TYPE	Neutral or slightly oxidizing
MATERIAL	1 pc. 3/32" × 3" × 6" carbon steel 1 pc. 3/32" × 1½" × 6" carbon steel		
PASSES Single	BEAD Weave	TIME (SEE INSTRUCTOR)	

Procedure 18
Horizontal T-Joint

OBJECTIVE

Upon completion of this lesson you should be able to braze-weld horizontal T-joints.

Text Reference: Section II, Lesson 6B; page 168.

PROCEDURE

1. Grind or sand the areas to be braze-welded.
2. Tack the two pieces together and position them as shown in Figure 18-1.
3. Heat both sides of the joint to a dull cherry color and add filler metal.
4. Use one of the weave motions you have learned to work the puddle to ensure it wets all the surfaces to the root of the fillet. Remember to use the flame to work distance to control the heat input.
5. Complete both sides of the joint, and clean and examine it.

42 OXYFUEL GAS WELDING AND BRAZING

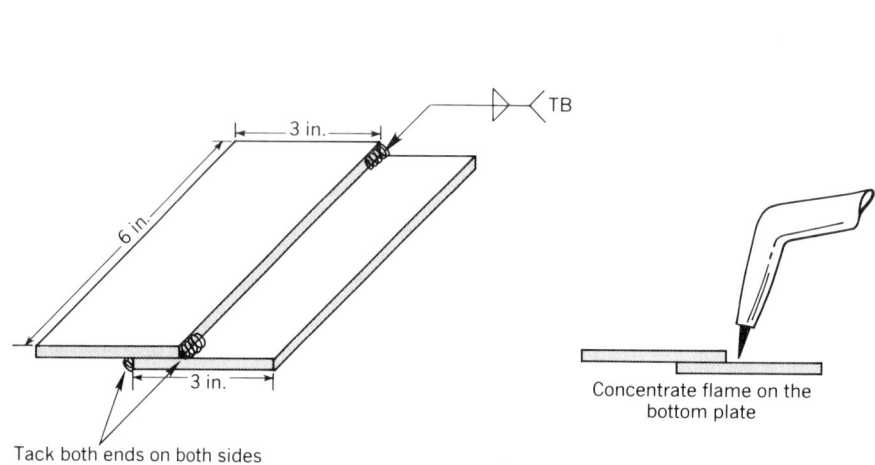

FIGURE 19-1

NAME			POSITION	
	Lap fillet			Horizontal
ELECTRODE		DIAMETER	FLAME TYPE	
Brazing rod		3/32"		Neutral or slightly oxidizing
MATERIAL				
	2 pcs. 3/32" × 3" × 6" carbon steel			
PASSES		BEAD	TIME (SEE INSTRUCTOR)	
Single		Weave		

Procedure 19

Horizontal Lap Joint Fillet

OBJECTIVE

Upon completion of this lesson you should be able to braze-weld horizontal lap fillets.

Text Reference: Section II, Lesson 6B; page 168.

PROCEDURE

1. Grind or sand the areas to be welded.
2. Tack the two pieces together and position as shown in Figure 19-1.
3. Heat the joint to a dull cherry color and add filler metal. Remember to concentrate the flame on the lower plate to balance the heat in the joint.
4. Use one of the weaving motions to work the puddle to ensure wetting on all areas of the joint.
5. Concentrate on keeping the fillet small, with the leg of the fillet not exceeding the thickness of the sheet. This will not be easy, as the flux will spread out and clean an area considerably larger than the immediate joint area. Use the flame to contain the filler metal to the immediate joint area.
6. Weld both sides of the joint and clean and examine it.

44 OXYFUEL GAS WELDING AND BRAZING

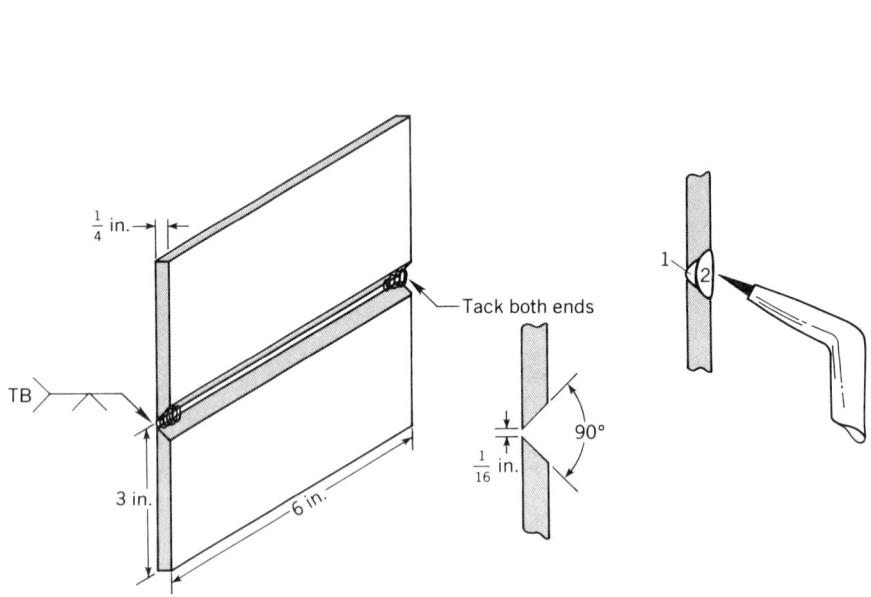

FIGURE 20-1

NAME		POSITION	
	V-groove butt		Horizontal
ELECTRODE	DIAMETER	FLAME TYPE	
Brazing rod	3/32"		Neutral or slightly oxidizing
MATERIAL			
2 pcs. ¼" × 3" × 6" carbon steel with a 30-degree bevel on one 6-in. side			
PASSES	BEAD	TIME (SEE INSTRUCTOR)	
Single	Weave		

Procedure 20

Horizontal V-Groove Butt Joint

OBJECTIVE

Upon completion of this lesson you should be able to braze-weld horizontal V-groove butt joints.

Text Reference: Section II, Lesson 6B; page 169.

PROCEDURE

1. Grind or sand the areas to be welded.
2. Tack the two pieces together and position as shown in Figure 20-1.
3. Heat the joint to a dull cherry color and add filler metal.
4. Use a weaving motion on the torch. Direct the flame toward the upper piece to direct the filler metal to the upper surface.

   Remember: The puddle will be very fluid and tend to run down toward the lower piece. Push the metal toward the top piece and withdraw the torch to allow the metal to cool and remain in place.

5. Do not attempt to complete the joint in one pass. Put several layers in, ending with a convex face surface with a smooth surface and even ripples.
6. Clean the plate and examine it. If you have properly spaced the root opening and paid attention to ensure filler metal wetting of all the joint surfaces, you will see a line of braze metal protruding through the root and joining both root surfaces.

46 OXYFUEL GAS WELDING AND BRAZING

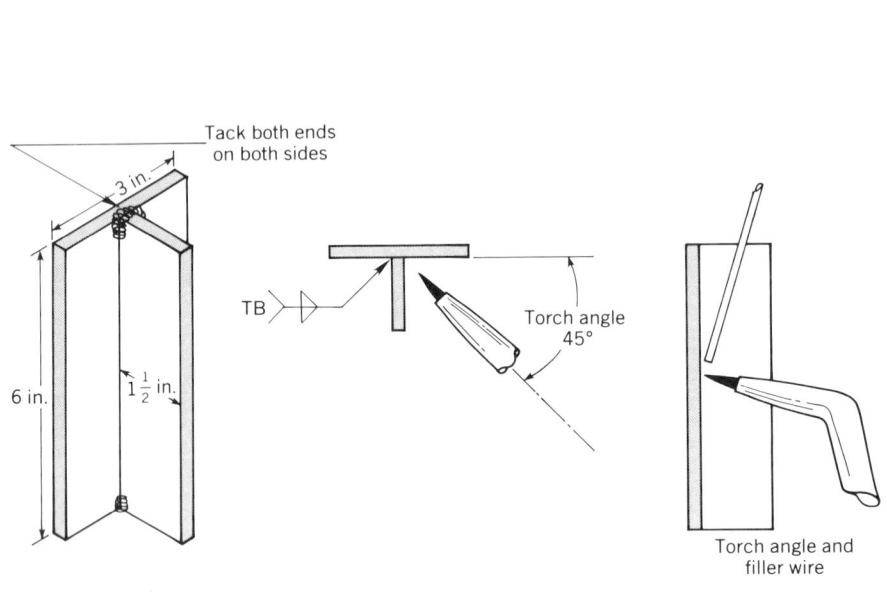

FIGURE 21-1

NAME	T-fillet	POSITION	Vertical
ELECTRODE Brazing rod	DIAMETER 3/32"	FLAME TYPE	Neutral or slightly oxidizing
MATERIAL	1 pc. 3/32" × 3" × 6" carbon steel 1 pc. 3/32" × 1 1/2" × 6" carbon steel		
PASSES Single	BEAD Weave	TIME (SEE INSTRUCTOR)	

Procedure 21
Vertical T-Joint Fillet

OBJECTIVE

Upon completion of this lesson you should be able to braze-weld vertical T-joint fillets.

Text Reference: Section II, Lesson 6C; page 170.

PROCEDURE

1. Grind or sand the areas to be braze-welded.
2. Tack the two pieces together and position as shown in Figure 21-1.
3. Heat both sides of the bottom of the joint and add filler metal. Make sure the joint is wet with the filler metal to the root.
4. Pull away the torch and allow the filler metal to freeze and form a shelf on which you can continue to build the vertical bead.
5. Observe that torch manipulation is very important in the vertical position. The puddle is very fluid and can easily become uncontrollable due to excessive heat. The torch flame has enough stiffness to move the puddle. Proper manipulation of the torch will turn this into an advantage. (See Figure 21-1.)
6. Complete the fillet on both sides. (It will be necessary to cool the joint between sides.) Clean the welds and examine them.

48 OXYFUEL GAS WELDING AND BRAZING

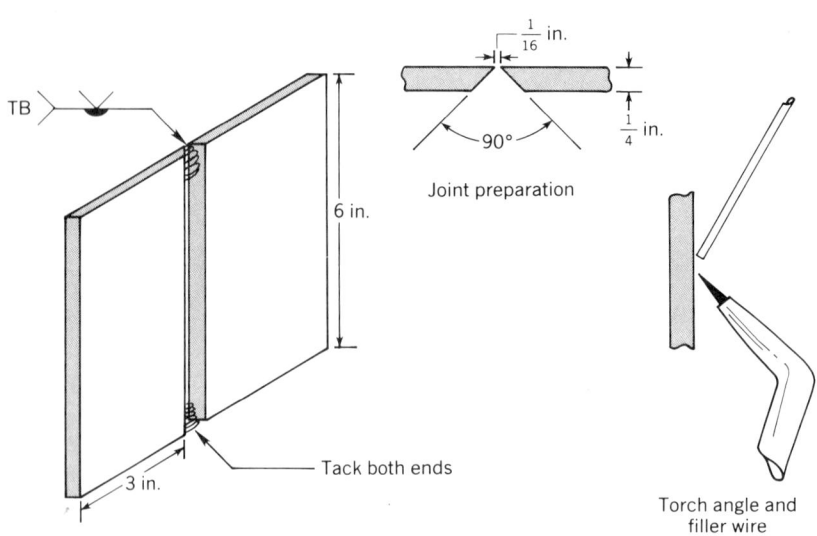

FIGURE 22-1

FIGURE 22-2

NAME			POSITION	
	V-butt			Vertical
ELECTRODE		DIAMETER	FLAME TYPE	
Brazing rod		3/32"		Oxidizing
MATERIAL				
2 pcs. ¼" × 3" × 6" carbon steel with a 30-degree bevel on one 6-in. side				
PASSES		BEAD	TIME (SEE INSTRUCTOR)	
Single		Weave		

Procedure 22

Vertical V-Groove Butt Joint

OBJECTIVE

Upon completion of this lesson you should be able to braze-weld vertical V-groove butt joints.

Text Reference: Section II, Lesson 6C; page 170.

PROCEDURE

1. Grind or sand the areas to be braze-welded.
2. Tack the two pieces together and position as shown in Figure 22-1.
3. Heat the joint until it is a dull cherry color and begin adding filler metal. Observe that all areas of the joint have been wet by the filler metal. This includes the groove faces and root.
4. Manipulate the torch to control the amount of heat put into the plate. One of the weaves you have learned will help to ensure the puddle covers the joint. Withdrawing the torch will help to keep the plate cool. If the plate becomes too hot the braze metal puddle will become too fluid and will fall from the joint. Use the shelf technique you have learned in the earlier lessons of this chapter.
5. Weld the joint in the two passes. (See Figure 22-2.) Pay particular attention to the side walls, or groove faces of the joint. If you keep the puddle too cool, or allow the weld face to become too convex, you may trap slag in these areas.
6. After completing the joint, clean and examine it.

Oxyacetylene Welding of Pipe

Pipe welders need special skills to weld pipe successfully. As a pipe welder you will need to change position continually to maintain the correct angle as you weld around a joint. This requires great flexibility. Good vision is needed to focus on the weld three dimensionally. Sometimes access to a particular pipe is limited and you may have to work with the aid of a mirror.

The procedures in this section will introduce you to techniques that will help you become a successful pipe welder.

GENERAL INSTRUCTIONS

1. Find a working stance that is both comfortable and secure; you will need both hands free to weld pipe.
2. Position yourself so that you have access to the joint and can move the torch and filler metal freely as you make the weld.
3. Maintain a relatively low-heat flame so that you can control the puddle.

MATERIALS AND EQUIPMENT

For this section the material and equipment will be as follows:

1. Welding goggles, number 5 or 6 lens
2. Safety glasses
3. Grinding goggles
4. Soapstone
5. Wraparound
6. Hand grinder or sander
7. Chipping hammer
8. Ballpeen hammer
9. Half-round bastard file, 10 in.
10. Wire brush

52 OXYFUEL GAS WELDING AND BRAZING

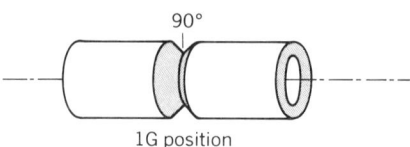

1G position

FIGURE 23-1

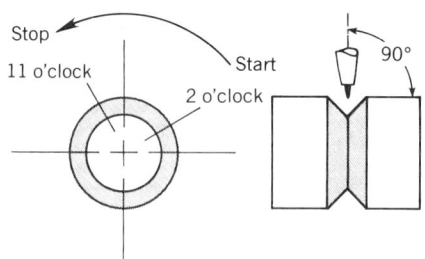

FIGURE 23-2

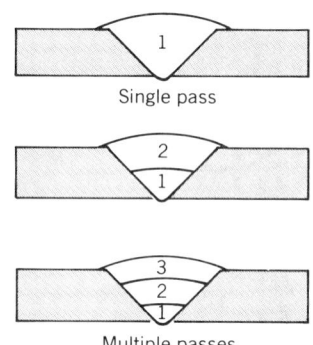

Single pass

Multiple passes

FIGURE 23-3

NAME	V-butt		POSITION	Flat
ELECTRODE Carbon steel	DIAMETER	3/32"	FLAME TYPE	Neutral
MATERIAL 2 pcs. pipe approximately 3 in. diameter × 3 in. long				
PASSES Single	BEAD	Weave	TIME (SEE INSTRUCTOR)	

Procedure 23

1G Position, Axis of the Pipe Horizontal

OBJECTIVE

Upon completion of this lesson you should be able to weld pipe in the flat 1G position.

Text Reference: Section II, Lesson 7A; page 172.

PROCEDURE

1. Tack the assembly together and position the pipe flat with the axis of the pipe horizontal. (See Figure 23-1.) Support the pipe so you can turn it for welding. You may support it by a piece of angle, or internally on a rod.
2. Weld the pipe from the 2 o'clock to the 11 o'clock position. (See Figure 23-2.) Establish a puddle before adding filler metal. Look for the keyhole, which will indicate root penetration.
3. When a keyhole has been established, begin adding filler metal. Use one of the weave techniques you have learned.
4. As you weld from 2 to 11 o'clock, you will notice that the bead will change in relation to its location on the pipe. At 2 o'clock the puddle will want to run downhill, at 12 o'clock the root will want to sag toward the inside of the pipe, and at 11 o'clock the puddle will again want to run downhill. To control the contour and width of the bead, you may try one or a combination of the following techniques.
    a. Alter the flame to pipe distance.
    b. Change the torch angle and use the flame to control the puddle.
5. When you reach the 11 o'clock position taper off your weld. Do this by moving the inner cone away from the crater and allowing it to solidify. Be careful to keep the crater in the flame's secondary envelope to prevent it from oxidation while it is cooling down.
6. Rotate the pipe to move the 11 o'clock position to a new position at 2 o'clock.
7. Bring the flame to a point ahead of the crater and reestablish a keyhole while melting the crater. When this is done begin welding toward the 11 o'clock position.
8. Repeat Steps 1 through 8 until the pipe is welded.
9. Depending on the diameter and wall thickness of the pipe, it may be necessary to use a multiple pass technique. See Figure 23-3 for examples of multiple pass methods.
10. When you are finished clean the pipe and examine it. You should see evidence of root penetration and root reinforcement around the entire inside circumference of the pipe. The outside surface or face of the weld should have an even reinforcement, a constant width, and a surface with fine ripples.

54 OXYFUEL GAS WELDING AND BRAZING

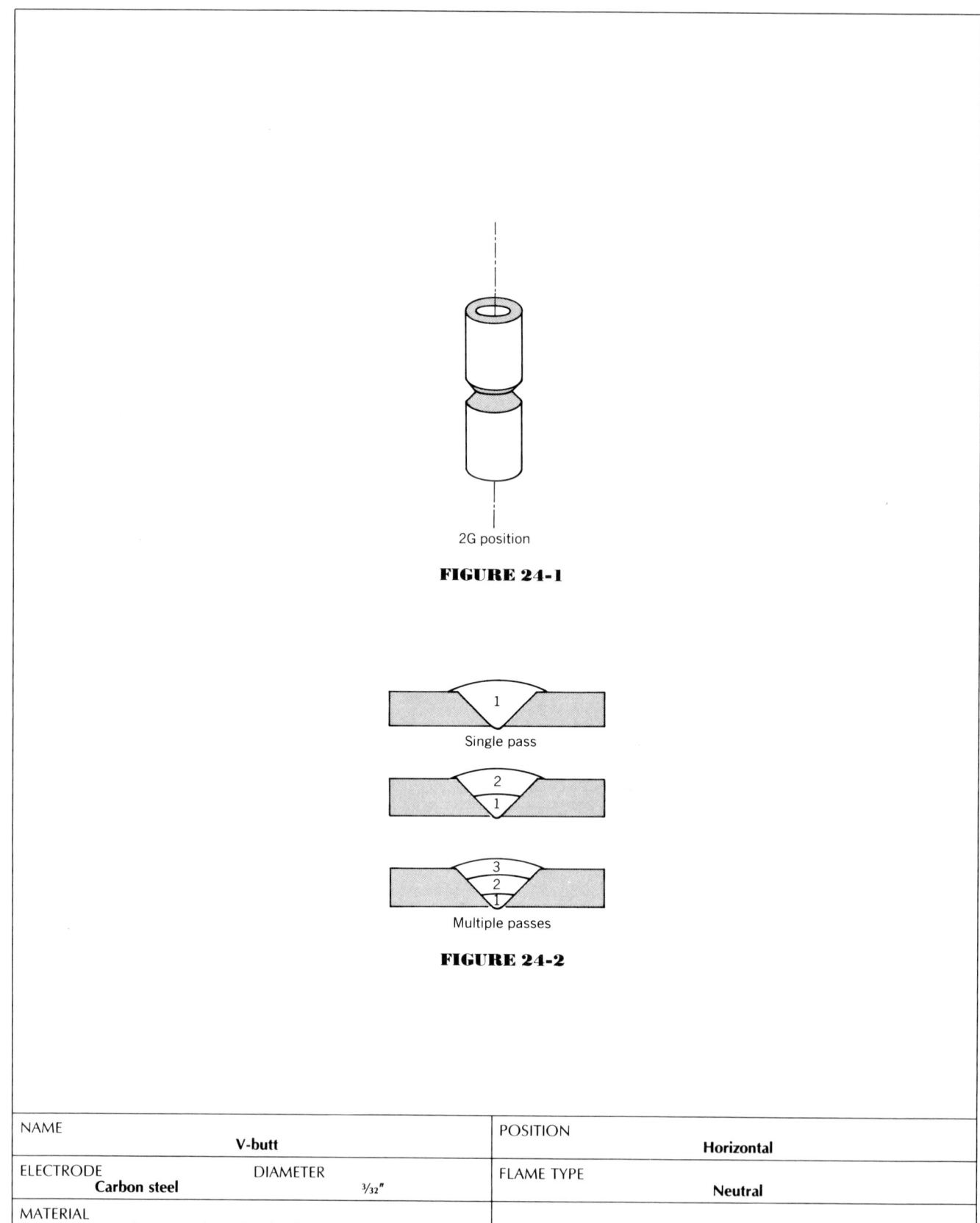

2G position

FIGURE 24-1

Single pass

Multiple passes

FIGURE 24-2

NAME V-butt		POSITION Horizontal	
ELECTRODE Carbon steel	DIAMETER 3/32"	FLAME TYPE Neutral	
MATERIAL 2 pcs. pipe approximately 3 in. diameter × 3 in. long			
PASSES Single	BEAD Weave	TIME (SEE INSTRUCTOR)	

Procedure 24

2G Position, Axis of the Pipe Vertical

OBJECTIVE

Upon completion of this lesson you should be able to weld pipe in the vertically fixed (2G) position.

Text Reference: Section II, Lesson 7A; page 173.

PROCEDURE

1. Tack two pieces of pipe approximately 3 in. diameter by 3 in. long, together and position as shown in Figure 24-1.
2. Begin the puddle using the keyhole technique to ensure proper root penetration. Add filler wire and begin the weld. Observe that as with the horizontal plate, groove weld, the puddle will want to sag. Build the weld from the bottom up, using the bottom as a shelf to support the rest of the weld.
3. During welding, you may rotate the pipe about its axis but you may not turn it end for end. Use the same technique you learned in the last lesson to start and stop your weld.
4. Depending on the wall thickness of the pipe you may use single or multiple pass techniques. (See Figure 24-2.)
5. When you have completed the weld, clean the pipe and examine it.

56 OXYFUEL GAS WELDING AND BRAZING

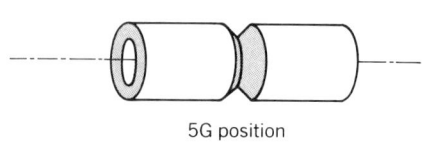

5G position

FIGURE 25-1

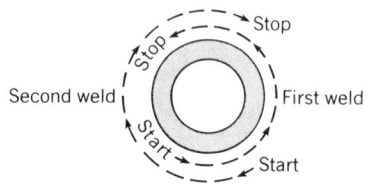

FIGURE 25-2

NAME	V-butt		POSITION	Horizontal-fixed
ELECTRODE Carbon steel	DIAMETER	3/32″	FLAME TYPE	Neutral
MATERIAL 2 pcs. pipe approximately 3 in. diameter × 3 in. long				
PASSES Single	BEAD	Weave	TIME (SEE INSTRUCTOR)	

Procedure 25

5G Position, Axis of the Pipe Horizontal

OBJECTIVE

Upon completion of this lesson you should be able to weld pipe in the horizontal fixed (5G) position.

Text Reference: Section II, Lesson 7A; page 174.

PROCEDURE

1. Tack two pieces of the pipe approximately 3 in. diameter by 3 in. long, together and position as shown in Figure 25-1.
2. To ensure a good weld in the 5G position the weld must be over lapped by going beyond the 6 and 12 o'clock position. (See Figure 25-2.) Begin the first weld at about the 7 o'clock position and begin welding vertical up around the pipe. Use the keyhole technique before adding filler metal.
3. Observe that the weld puddle will act differently in relation to its position on the pipe. Use one or a combination of the techniques you learned previously to control the puddle.
4. When stopping and restarting the welds be sure to use the methods you learned earlier. It will be necessary to stop and start frequently to reposition yourself and the torch.
5. Depending on the thickness of the pipe it may be necessary to use a multiple pass technique.
6. When you have finished the weld, clean and examine it.

SHIELDED METAL ARC WELDING PROCESS

Shielded Metal Arc Welding of Plate

Shielded metal arc welding is a welding process widely used in industries involving manufacturing with metals. The skills you develop in shielded metal arc welding will serve as a foundation upon which you can build a mastery of more complex welding procedures.

Shield metal arc welding, or *stick welding,* involves the use of electrodes. Electrodes are manufactured to weld a wide range of metals. The Procedures in this section will guide you through the steps in welding carbon steel using various electrodes in all of the common welding positions. A number of electrodes are specified so that you become familiar with their characteristics and gain experience in handling them.

GENERAL INSTRUCTIONS

Throughout the procedures in this section you may use either an AC or DC welder, depending on their availability. When a DC welder is used, the polarity will be indicated on the procedure sheet.

The lessons have been written for the right-handed student. However, left-handed students should not have any trouble. They should start at the opposite end of a plate and reverse the right-hand electrode drag and push angles.

All welders should learn to weld both from right to left and left to right. In addition, they should also learn to weld toward and away from themselves. Proficient welders can weld in any direction.

MATERIALS AND EQUIPMENT

For this section the material and equipment will be as follows:

1. Welding shield
2. Safety glasses
3. Protective leathers and gloves
4. Chipping hammer and wire brush

62 SHIELDED METAL ARC WELDING PROCESS

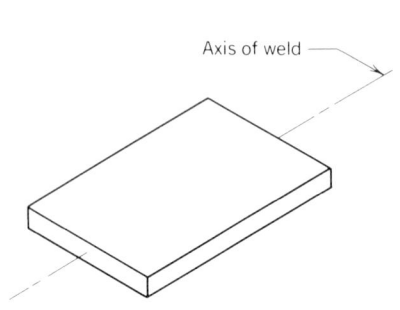

FIGURE 26-1

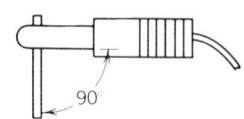

FIGURE 26-2

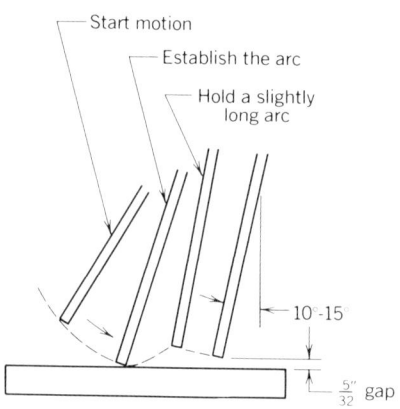

FIGURE 26-3

NAME	Striking an arc	POSITION	Flat
ELECTRODE E-6010 or E-6011	DIAMETER 5/32" 5/32"	AMPERAGE	110-170 DCEP 110-170 AC or DC
MATERIAL	1 pc. 1/4" × 4" × 8" carbon steel		
PASSES None	BEAD None	TIME (SEE INSTRUCTOR)	

Procedure 26

Striking and Maintaining an Arc

OBJECTIVE

Upon completion of this lesson you should be able to strike and maintain an arc.

Text Reference: Section III, Lesson 5A; page 219.

PROCEDURE

1. Wire brush the welding surface of the plate. The axis of the weld will be as shown in Figure 26-1.
2. Secure the workpiece clamp and adjust the welding current. DC current only can be used with E-6010 electrode. Either AC or DC can be used with E-6011 electrode. This statement holds true throughout the book.
3. Place the electrode in the holder at a 90-degree angle. (See Figure 26-2.)
4. Hold the electrode at approximately a 30-degree angle off the perpendicular about 1 in. above the left-hand edge of the plate.
5. Lower your shield, then bring the electrode down and to the right using a swinging motion, like striking a match. (See Figure 26-3.)
6. When the electrode contacts the plate, it will strike an arc. When this happens, raise the electrode slightly to about 3/16 in. (See Figure 26-3).
7. Immediately lower the electrode to reduce the arc gap. The distance between the tip of the electrode and the workpiece should be approximately the thickness of the electrode core wire, 3/32 in. (See Figure 26-3).
8. Do not lift your shield, but break the arc by snapping the electrode up to the left. The power supply cannot provide the correct arc current to a long arc. As you increase the arc length the current is decreased. There will be some arc length that causes the arc to extinguish suddenly. When this happens, restart the arc exactly as you did the first time. Continue to strike, and strike the arc until the electrode burns down to approximately 2 in.
9. The electrode melts with each start. It gets shorter and you must constantly feed the electrode toward the puddle. If you do not lower the electrode, the arc gap will increase, and the arc will go out.

64 SHIELDED METAL ARC WELDING PROCESS

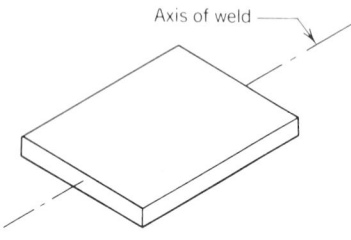

FIGURE 27-1

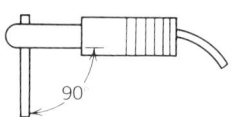

FIGURE 27-2

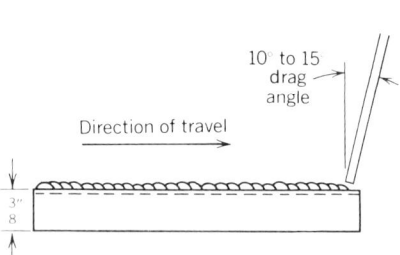

FIGURE 27-3

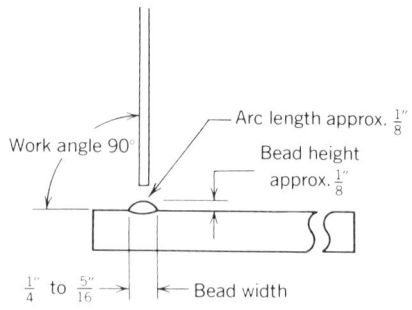

FIGURE 27-4

NAME	Welding stringer beads	POSITION	Flat
ELECTRODE E-6010 or E-6011	DIAMETER 1/8" 1/8"	AMPERAGE	75-110 DCEP 75-110 AC or DC
MATERIAL	1 pc. 3/8" × 4" × 8" carbon steel		
PASSES None	BEAD None	TIME (SEE INSTRUCTOR)	

Procedure 27

Welding Stringer Beads

OBJECTIVE

Upon completion of this lesson you should be able to weld flat stringer beads.

Text Reference: Section III, Lesson 5A; page 220.

PROCEDURE

1. Wire brush the welding surface of the plate.
2. The plate must be welded with the axis as shown in Figure 27-1.
3. Attach the workpiece clamp and adjust the power supply as directed.
4. Use a clamp to hold your workpiece in place.
5. Make sure the workpiece cable is securely attached.
6. Place the electrode in the holder at a 90-degree angle. (See Figure 27-2.)
7. Assume a comfortable position, with your left elbow resting on the table. Place your right hand, the one with the electrode holder, in the palm of your left hand. Do not grip the holder tightly. If you do, you will tire quickly and your hand will become unsteady.
8. Strike the arc on the left side of the plate. Tilt the electrode holder slightly to the right. The electrode should point at about a 10 to 15 degree angle from the vertical into the puddle. This is commonly called the lead or drag angle. (See Figure 27-3.)
9. Maintain an arc length of about 1/8 in. and keep the width of the puddle to 2 to 2 1/2 times the diameter of the core wire and approximately 3/32 in. in height to 5/16 in. wide. (See Figure 27-4.)
10. Run the stringer bead for about 3 in.

Note: Examine the stringer for evenness of the ripples, width, and height. Chip the slag away and wire brush thoroughly before continuing.

11. Strike another arc about 3/4 in. to the right of the weld puddle crater at the end of the stringer. Then move the arc directly over the crater. It should be a slightly long arc. Hold it in position for a moment. Then close the arc gap to normal and proceed as before.
12. Continue in the same manner, but start the next stringer 1/2 in. below the first. Completely cover the plate.

66 SHIELDED METAL ARC WELDING PROCESS

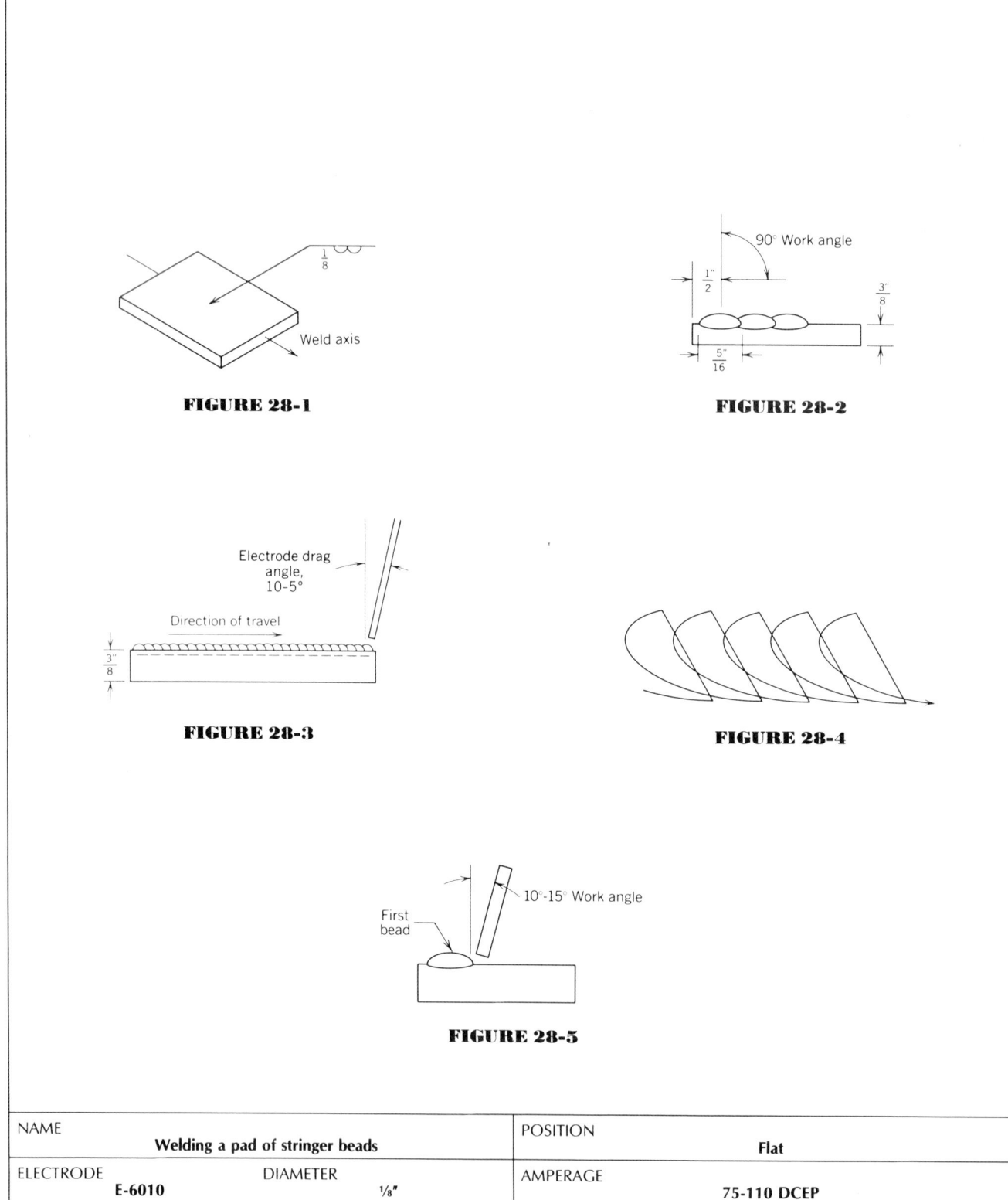

FIGURE 28-1

FIGURE 28-2

FIGURE 28-3

FIGURE 28-4

FIGURE 28-5

NAME	Welding a pad of stringer beads	POSITION	Flat
ELECTRODE	DIAMETER	AMPERAGE	
E-6010	1/8"		75-110 DCEP
or			
E-6011	1/8"		75-110 AC or DC
MATERIAL	1 pc. 3/8" × 4" × 10" carbon steel		
PASSES Multiple	BEAD Stringer	TIME (SEE INSTRUCTOR)	

Procedure 28

Running a Flat Pad of Stringers

OBJECTIVE

Upon completion of this lesson you should be able to run a flat pad of stringers.

Text Reference: Section III, Lesson 5A; page 223.

PROCEDURE

1. Wire brush the welding surface of the plate.
2. The plate must be welded with the axis as shown in Figure 28-1.
3. Secure the workpiece and the workpiece clamp.
4. Adjust the welding current.
5. Make sure that the electrode is held firm in the electrode holder, at a 90-degree angle.
6. Start at the left side of the plate, about ½ in. from the top edge. (See Figure 28-2.) Use about a 10 to 15 degree drag angle. (See Figure 28-3.)
7. Strike and hold a slightly long arc until a puddle begins to form. Then shorten the arc length and move steadily to the right. Use a "C" motion. (See Figure 28-4.) With a "C" motion the bottom swing of the electrode moves further to the right than the top portion. Pause slightly at the top of the "C." Then move down, out and back somewhat faster.

Note: Do not attempt to use the "C" motion until the instructor demonstrates it.

8. Clean the stringer thoroughly using the chipping hammer and the wire brush.
9. The next bead should overlap about one-third to one-half of the first bead. Use a drag angle of 10 to 15 degrees.
10. Use the angles mentioned above for this bead and all the remaining beads in this assignment. Strike the arc at the left side of the plate. Hold the electrode so that the edge of the coating that is farthest from you is pointed at the nearest edge of the first bead. (See Figure 28-5.)
11. In Figure 28-5, notice the angle of the electrode with relation to first bead and the surface of the plate.
12. Strike and hold a long arc. Then proceed along the closest edge of the first bead. Make sure that you limit the motion of the electrode. The stringer should not be wider than 2 to 2½ times the diameter of the core wire.
13. The second stringer must cover about one-third to one-half of the first one, as shown in Figure 28-2.
14. Continue to weld stringer beads in this manner. Cover the entire plate. Then cool the plate and examine it.

68 SHIELDED METAL ARC WELDING PROCESS

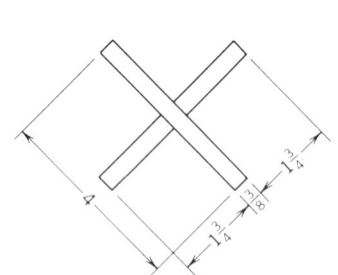

FIGURE 29-1

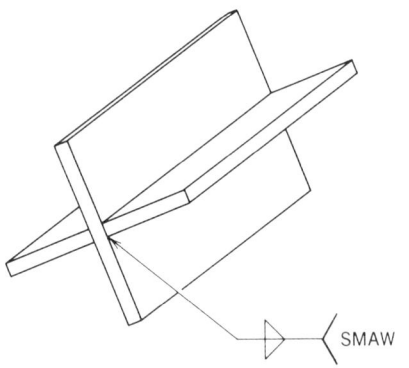

FIGURE 29-2

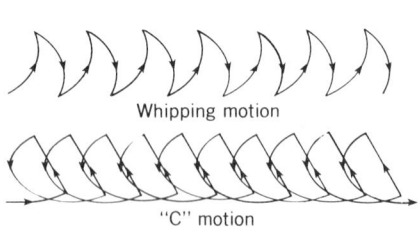

FIGURE 29-3

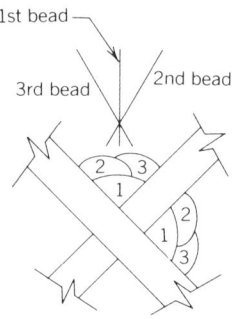

FIGURE 29-4

NAME	T-fillet	POSITION	Flat
ELECTRODE E-6010 or E-6011	DIAMETER 1/8" 1/8"	AMPERAGE	75-110 DCEP 75-110 AC or DC
MATERIAL	1 pc. 3/8" × 4" × 12" carbon steel 2 pcs. 3/8" × 1 3/4" × 12" carbon steel		
PASSES Multiple	BEAD Stringer	TIME (SEE INSTRUCTOR)	

Procedure 29

T-Joint Fillet, 1F Position, Stringer Beads (E-6010 or E-6011 Electrodes)

OBJECTIVE

Upon completion of this lesson you should be able to weld flat fillets.

Text Reference: Section III, Lesson 5A, page 224.

PROCEDURE

1. Remove all slag, rust, or mill scale from the weld surface of the plates.
2. Tack the three plates as shown in Figure 29-1. Be sure that there is little or no space between surfaces. Make sure that the plates are perpendicular to each other.
3. Position the practice piece as shown in Figure 29-2.
4. Be sure that the workpiece and the work lead are secure.
5. With the electrode perpendicular to the joint run a stringer from left to right. Maintain an electrode drag angle of 10 to 15 degrees. Use either a whipping motion or the "C" motion as in Figure 29-3.
6. Chip and wire brush the bead thoroughly. Cool the plate. Then inspect it for undercut and unevenness of height or width. Undercut may be caused by too long an arc, too high a current, wrong electrode angle or manipulation, or too fast a rate of travel.
7. Overlap a second stringer bead on the first one. Run half on the first bead and half on the side of the workpiece farthest from you. Use a 10 to 15 degree angle and a 15 to 20 degree work angle. This will cause the metal to flow up the side of the workpiece. One-half of the bead should be on the plate and the other half covering about one-half of the first bead. (See Figure 29-4.) Using this angle you will find that the molten metal flows or washes up the side of the plate that the electrode is pointed toward.
8. Clean the weld thoroughly. Run another stringer with the electrode pointed toward you, at about a 15 to 20 degree angle. This bead should cover about one-half of the last stringer and flow up the side of the plate. The complete weld should be flat or slightly convex and have legs of equal size.
9. Repeat the task, this time use the joint opposite the one you just welded.
10. Complete two of the joints and save two for the next lesson.

70 SHIELDED METAL ARC WELDING PROCESS

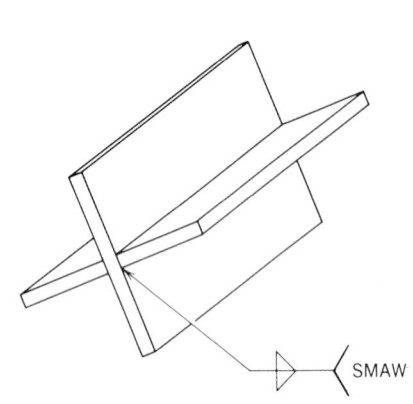

FIGURE 30-1

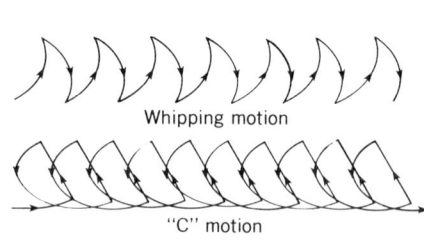

FIGURE 30-2

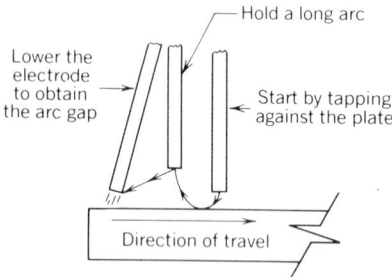

FIGURE 30-3

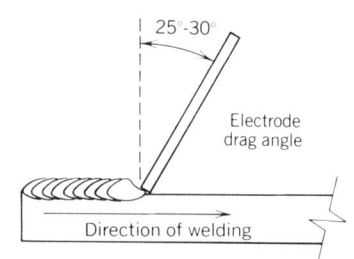

FIGURE 30-4

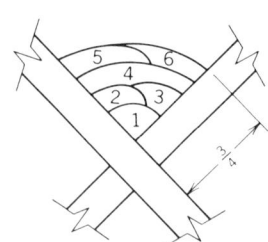

FIGURE 30-5

NAME	T-fillet	POSITION	Flat
ELECTRODE	DIAMETER	AMPERAGE	
E-7014 or E-7024	1/8" 5/32" 1/8" 5/32"		110-160 AC or DCEP 150-210 110-160 AC or DC 150-210
MATERIAL	Use the two joints left over from Procedure 29		
PASSES Multiple	BEAD Stringer	TIME (SEE INSTRUCTOR)	

Procedure 30

T-Joint Fillet, 1F Position, Stringer Beads (E-7014 or E-7024 Electrodes)

OBJECTIVE

Upon completion of this lesson you should be able to weld a pad of stringer beads in the 1F position.

Text Reference: Section III, Lesson 5A; page 225.

PROCEDURE

1. Wire brush the surfaces to be welded.
2. Position the practice piece as shown in Figure 30-1.
3. Make sure the workpiece clamp and the workpiece are secure.
4. Adjust the welding current.
5. Weld the first joint with E-7014 electrodes. Use a drag angle of 10 to 15 degrees, as you did with the E-6010 and E-6011 electrodes. But you do not need to use the "C" motion (see Figure 30-2); just move the electrode along the joint without any movement other than forward travel.
6. After each pass chip and wire brush thoroughly. Although slag from the E-7014 electrode comes off easily, it often sticks to tiny crevices, especially in the corners or toes of the weld.
7. Pay close attention to the edges or toes of the weld. Make sure you are not leaving any undercut. If you are, slow down. Use a lower rate of travel. Make sure you are not holding too long an arc.
8. Place the second joint in the same position as the first. Use the E-7024 electrodes.
9. The E-7024 electrode is a heavily coated iron powder electrode. Because of this, it can be dragged along the joint with the coating touching the metal being welded. This prevents it from stubbing out.
10. Drag the first pass along the root of the joint. Keep the electrode perpendicular to the workpiece. Use a drag angle of approximately 30 degrees. Strike the arc by tapping the tip of the electrode in the groove of the workpiece to break the coating off the tip. (See Figure 30-3.)
11. When the arc starts, hold a long arc for a moment. Then shorten the arc length until the coating on the right-hand side of the electrode touches the metal, at a 30-degree lead angle. (See Figure 30-4.) Weld at a travel speed that gives the proper bead width for the electrode diameter used.
12. Be sure you maintain the correct drag angle. If you reduce the drag angle too much, the electrode can stub out.
13. Continue, as with the E-7014 electrode, using the same side angles.
14. After you have welded the first three stringers, cover over the weld with a weave bead. Increase the arc current setting and with the same electrode angles strike the arc on the left. Then move to the right, weaving the electrode from side to side. (See Figure 30-2.) Pause slightly at the sides to minimize undercut.
15. Part of the next weave bead should cover slightly more than half of the previous bead, while the rest of the bead washes up the side of the joint. (See Figure 30-5.)
16. Part of the next weave bead should cover about one-third of the previous bead, while the rest of the bead washes up the opposite side of the joint. (See Figure 30-5.) Continue to run beads in this manner. Make sure they are laced together, the same as you did with stringer beads.

72 SHIELDED METAL ARC WELDING PROCESS

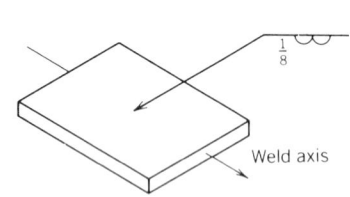

FIGURE 31-1

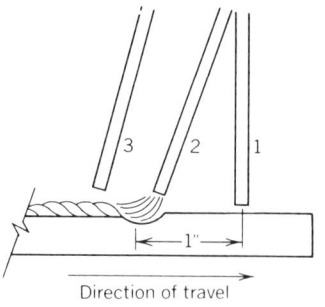

FIGURE 31-2

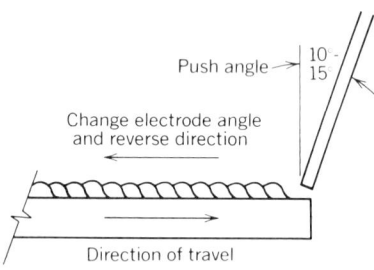

FIGURE 31-3

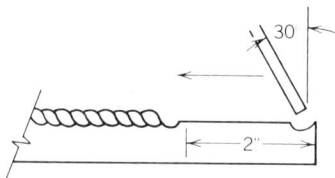

FIGURE 31-4

NAME			POSITION	
	Filling craters			Flat
ELECTRODE		DIAMETER	AMPERAGE	
	E-6010	⅛"		75-125 DCEP
	or			
	E-6011	⅛"		75-125 AC or DC
MATERIAL				
	1 pc. ¼" × 4" × 12" carbon steel			
PASSES		BEAD	TIME (SEE INSTRUCTOR)	
	Multiple	Stringer		

Procedure 31

Filling Craters at the End of Welds

OBJECTIVE

Upon completion of this lesson you should be able to fill craters properly.

Text Reference: Section III, Lesson 5A; page 226.

PROCEDURE

1. Wire brush the surface of the plate.
2. Position the plate as shown in Figure 31-1.
3. Secure the plate and attach the workpiece clamp.
4. Adjust the welding current.
5. Start at the left. Run a short stringer, about 2 or 3 in. Stop. Then clean the weld.
6. Strike another arc about 1 in. ahead of the crater. Then move the arc on top of the crater. See Figure 31-2. Whenever you strike an arc, it leaves a hard spot in the metal. When you begin ahead of the crater, the hard spot is covered over by the bead and is eliminated.
7. Hold a normal arc on the crater until it is filled to the same height as the bead. Then continue the stringer. Stop every few inches and repeat the process on the new crater. This will give you practice with this method.
8. When you reach the end of the plate, reverse the direction of travel. Raise the electrode slowly to keep the same arc length as you weld up and over the finished bead. Weld back over the bead for ½ in. Hold a close arc, then stop welding. Use a 10- to 15-degree push angle. (See Figure 31-3.) If you burn a hole though the weld, or the weld metal spills over and runs off the end of the plate, you either moved too slowly or carried too long an arc. Practice this method until you can do it well.
9. Another method of crater filling is to break the arc about 2 in. from the end of the plate. Clean the bead. Then start at the right side of the plate and weld back toward the crater with the electrode pointing into the puddle at a drag angle of about 15 to 30 degrees. (See Figure 31-4.) Weld over the crater of the bead and break the arc slowly or with a slight circular motion.

74 SHIELDED METAL ARC WELDING PROCESS

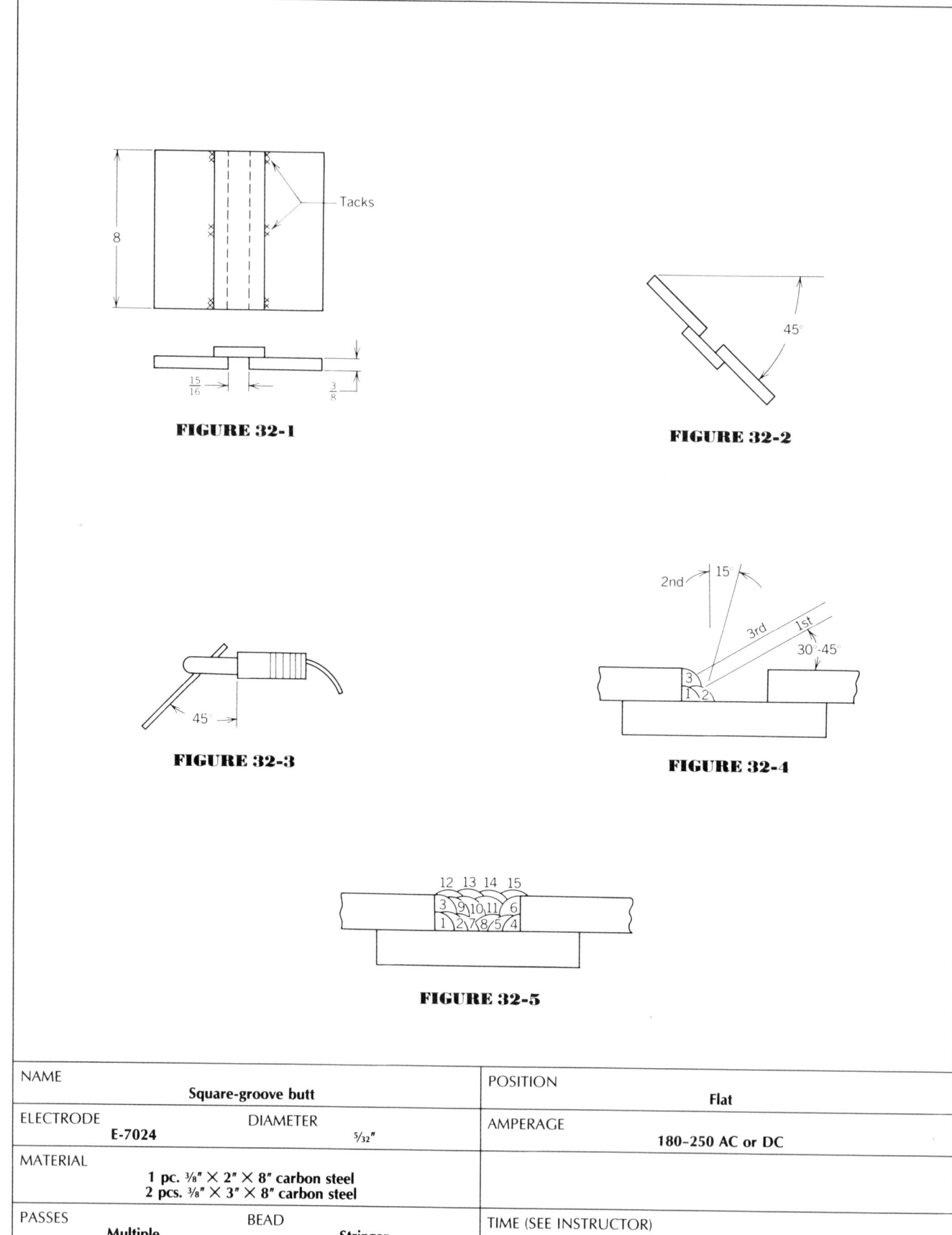

NAME	Square-groove butt	POSITION	Flat
ELECTRODE E-7024	DIAMETER 5/32"	AMPERAGE	180–250 AC or DC
MATERIAL	1 pc. 3/8" × 2" × 8" carbon steel 2 pcs. 3/8" × 3" × 8" carbon steel		
PASSES Multiple	BEAD Stringer	TIME (SEE INSTRUCTOR)	

Procedure 32

Square-Groove Butt Joint, 1G Position with Backup Bar, Stringer Beads

OBJECTIVE

Upon completion of this lesson you should be able to weld square-groove butt joints with a backup bar in the flat position.

Text Reference: Section III, Lesson 5A; page 227.

PROCEDURE

1. Clean all slag, rust, mill scale, and other dirt from the parts to be welded.
2. Tack weld the pieces as shown in 32-1. It is important that you have as good a fit as possible. Be sure that there are no spaces between the parts.
3. Position the test piece as shown in Figure 32-2.
4. Make sure the work clamp and the test piece are securely fastened.
5. Adjust the welding current.
6. Make sure your tack welds are at least ¾ in. long and are placed according to Figure 32-1. If the tacks are not strong enough or in the proper places, the plates may distort and make the test more difficult to pass.
7. Place the electrode in the holder at a 45-degree angle. (See Figure 32-3.)
8. Fill the two fillets with three stringer beads. (See Figure 32-4.)
9. Hold the electrode at approximately 30 to 45 degrees from the surface of the plate with a drag angle of about 30 degrees. Drag it along the bottom leg of the joint. The electrode coating should touch both the backup bar and the vertical leg. (See Figure 32-4.)
10. Move along steadily. Make sure the bead covers both legs of the joint equally.
11. Clean and run the second bead with the electrode. Hold the electrode about 15 degrees off the perpendicular as in Figure 32-4.
12. Remove the slag from the weld deposit. Run the third bead with the same electrode angle used with the first bead. (See Figure 32-4.) Pay close attention to the top, or leading edge of the upper plate. Watch the top edge of the puddle. Make sure that you do not cut down the top edge. The fillet is not complete if you do not cover the entire vertical leg.
13. Weld the other fillet in the same manner.
14. Fill in the remainder of the joint with either the stringer bead or the weave method. (See Figure 32-5.)

76 SHIELDED METAL ARC WELDING PROCESS

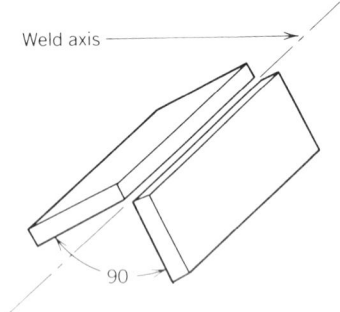

FIGURE 33-1

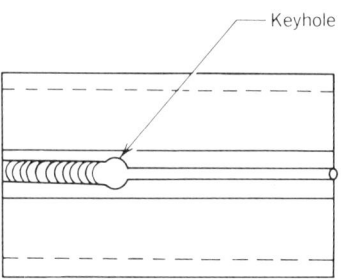

FIGURE 33-2

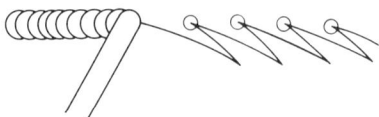

FIGURE 33-3

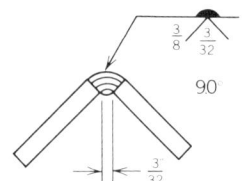

FIGURE 33-4

NAME		POSITION	
	Outside corner		Flat
ELECTRODE	DIAMETER	AMPERAGE	
E-6010	1/8"		75–125 DCEP
or			
E-6011	1/8"		75–125 AC or DC
MATERIAL			
2 pcs. 3/8" × 2" × 12" carbon steel			
2 spacers, 3/32" × 3/4" × 5" long			
PASSES	BEAD	TIME (SEE INSTRUCTOR)	
Multiple	Stringer and weave		

Procedure 33

Outside-Corner Joint, 1G Position, Stringer and Weave Beads

OBJECTIVE

Upon completion of this lesson you should be able to weld open root outside-corner joints in the flat position using stringer and weave beads.

Text Reference: Section III, Lesson 5A; page 228.

PROCEDURE

1. Clean all the parts to be welded.
2. Use the spacers to position the pieces. Then tack weld the two pieces together to form a 90-degree angle as in Figure 33-1. Be sure the space along the entire joint is even.
3. Remove the spacers and place the weldment in position. (See Figure 33-1.)
4. Make sure the workpiece clamp and workpiece are securely fastened. Adjust the welding current.
5. The tacks must be at least 1 in. long or the root opening may close up as you weld the first pass.
6. Hold the electrode perpendicular and use a drag angle of 10 to 15 degrees. Weld over the tack at the left side of the joint. When you reach the end of the tack, pause until a keyhole appears. (See Figure 33-2.)
7. When the keyhole appears, whip the electrode foreward and slightly up one side of the joint. (See Figure 33-3.) This motion of whipping out and away from the puddle allows the molten metal to cool and freeze. Otherwise it would fall through the root and cause excessive melt through.
8. As the filler metal solidifies, return the electrode to the forward edge of the keyhole. Hold it there until it fills the keyhole and another keyhole opens. Continue whipping the arc across the entire joint, adding a dab of metal each time.
9. The second pass should be a stringer bead. Make sure you fuse both sides of the joint at the toes into the edges of the weld. (See Figure 33-4.)
10. The third pass is a weave bead. Pause slightly at the sides for proper fusion. (See Figure 33-3.)
11. The final pass is also a weave bead. Use the motion shown in Figure 33-3.
12. Watch out for excessive overlap. Avoid extra weld metal hanging over the edge of the joint.
13. The completed weld should have a slightly convex (outwardly curved) appearance.

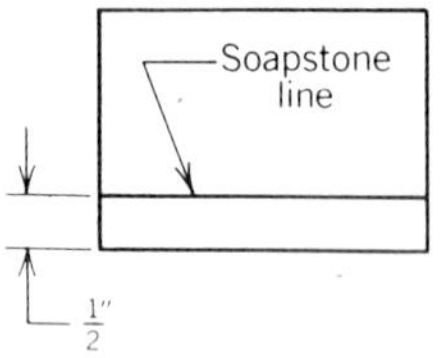

78 SHIELDED METAL ARC WELDING PROCESS

FIGURE 34-1

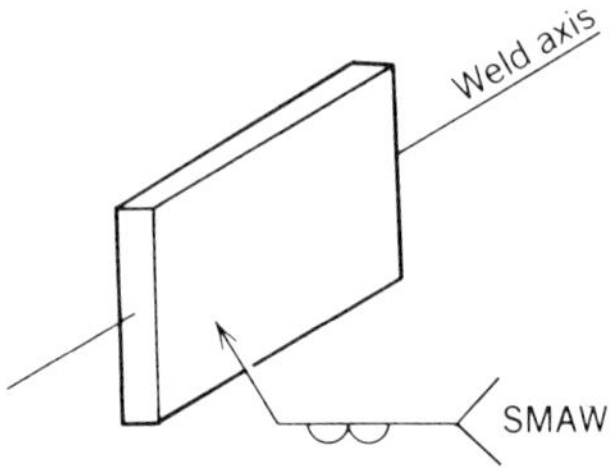

FIGURE 34-2

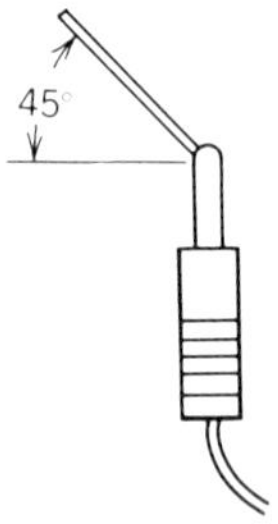

FIGURE 34-3

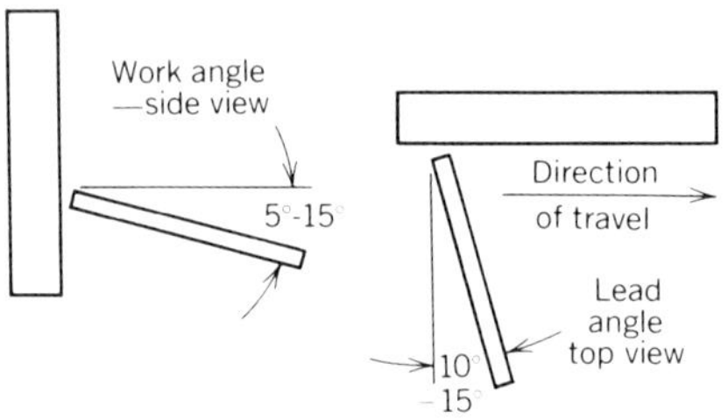

FIGURE 34-4

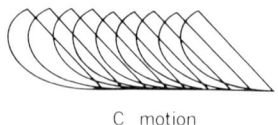
Whipping motion.

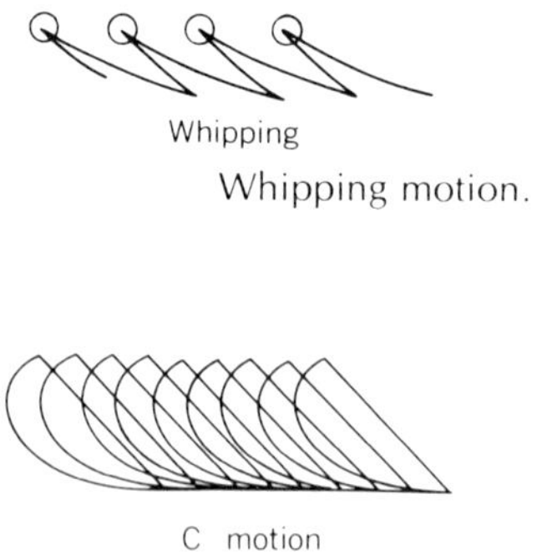

C motion

FIGURE 34-5

NAME	Pad of stringers		POSITION	Horizontal
ELECTRODE	DIAMETER		AMPERAGE	
E-6010		1/8"		75–125 DCEP
or **E-6011**		1/8"		75–125 AC or DC
MATERIAL	1 pc. 3/8" × 6" × 6" carbon steel			
PASSES	BEAD		TIME (SEE INSTRUCTOR)	
Multiple		**Weave**		

Procedure 34

Running a Horizontal Pad of Stringers

OBJECTIVE

Upon completion of this lesson you should be able to weld a pad of stringers in the horizontal position.

Text Reference: Section III, Lesson 5B; page 230.

PROCEDURE

1. Be sure the edges of the plate are clean, then wire brush the surface of the plate.
2. Draw a line across the plate with the soapstone. Place the line ½ in. from the bottom edge of the plate as in Figure 34-1.
3. Position the plate as in Figure 34-2 with the line at the bottom.
4. Adjust the welding current.
5. Place the electrode in the holder. Clamp it at a 45-degree angle as shown in Figure 34-3.
6. Start at the left side of the plate. Hold the electrode pointing upward at about a 5- to 15-degree angle. Use a 10- to 15-degree drag angle as shown in Figure 34-4.
7. Move to the right with a slight whipping motion. You can also use the "C" motion. Each of these motions gives the molten puddle a chance to cool slightly. This helps it solidify and keeps it from sagging. When using the "C" motion, do not forget to pause at the upper left part of the "C." (See Figure 34-5.)
8. The second stringer, and all that follow, should cover the top third of the bead below it. This keeps the weld surface even in height, without valleys or low spots.

80 SHIELDED METAL ARC WELDING PROCESS

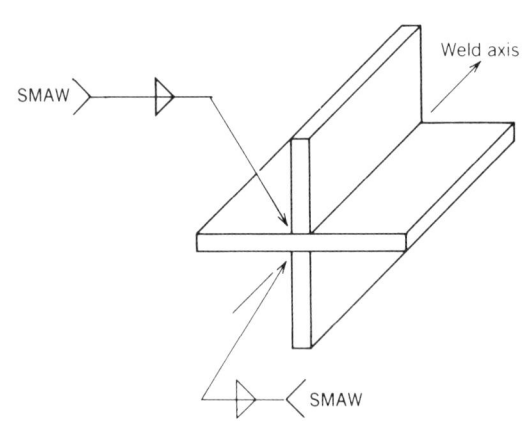

FIGURE 35-1

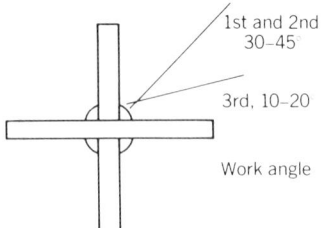

FIGURE 35-2

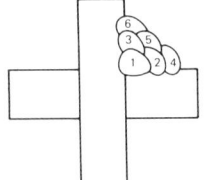

FIGURE 35-3

NAME	Fillet		POSITION	Horizontal
ELECTRODE	DIAMETER		AMPERAGE	
E-6010		1/8"		75-125 DCEP
or				
E-6011		1/8"		75-125 AC or DC
MATERIAL	1 pc. 3/8" × 4" × 12" carbon steel			
	2 pcs. 3/8" × 1 1/2" × 12" carbon steel			
PASSES	BEAD		TIME (SEE INSTRUCTOR)	
Multiple	Weave			

Procedure 35

T-Joint Fillet, 2F Position, Multipass Stringers (E-6010 or E-6011 Electrodes)

OBJECTIVE

Upon completion of this lesson you should be able to weld horizontal fillets.

Text Reference: Section III, Lesson 5B, page 230.

PROCEDURE

1. Remove all slag and other impurities from the plates. Clean the area to be welded.
2. Tack the plates to form a cross as shown in Figure 35-1. Do not hold the metal in place with your gloves. Use some scrap metal or a clamp to position the plates.
3. Position the weldment as in Figure 35-1.
4. Adjust the welding current.
5. Place the electrode in the holder with a 45-degree angle.
6. Hold the electrode at about a 30- to 45-degree angle off the bottom plate as in Figure 35-2. Use a 5- to 15-degree drag angle for the first pass. Fill the crater at the end of the bead.
7. Hold the same electrode angles with the second bead. Point the tip of the electrode so it is partially on the plate and partially on the first bead as shown in Figure 35-2. Slightly more than half of the filler metal should be on the stringer bead.
8. For the third stringer lower the electrode angle to between 10 and 20 degrees, as in Figure 35-2. Use a whip or "C" motion. Undercut can be reduced if you whip properly or pause at the upper left side of the puddle when using the "C" motion.
9. Clean and inspect the last bead for undercut. Weld three more stringers over the first three passes to complete the joint. See the bead sequence in Figure 35-3.
10. Clean your weldment and inspect it. Then weld one more joint.

82 SHIELDED METAL ARC WELDING PROCESS

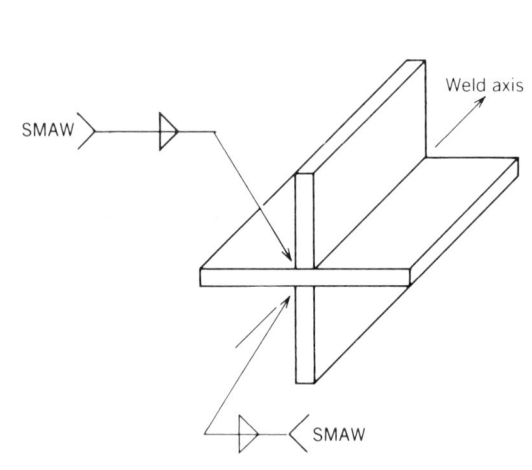

FIGURE 36-1

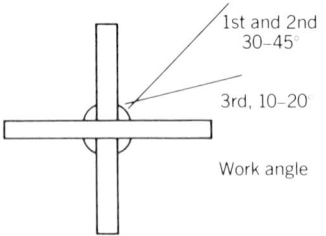

FIGURE 36-2

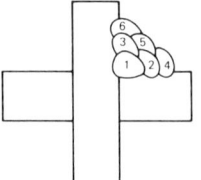

FIGURE 36-3

NAME			POSITION	
	Fillet			**Horizontal**
ELECTRODE		DIAMETER	AMPERAGE	
E-7024		**1/8"**		**140–190 AC or DCEP**
MATERIAL				
Use the two joints left over from Procedure 35				
PASSES		BEAD	TIME (SEE INSTRUCTOR)	
Multiple		**String**		

Procedure 36

T-Joint Fillet, 2F Position, Multipass Stringers (E-7024 Electrode)

OBJECTIVE

Upon completion of this lesson you should be able to weld horizontal fillets with iron powder electrodes.

Text Reference: Section III, Lesson 5B; page 231.

PROCEDURE

1. Position the assembly in the position shown in Figure 36-1.
2. Place the E-7024 electrode in the holder at a 45-degree angle.
3. Adjust the welding current.
4. Run a stringer bead from left to right. Use a drag technique. Hold the electrode at about 30- to 45-degrees to the bottom plate, (see Figure 36-2), with a drag angle of about 30 degrees. Stop welding at the center of the joint—clean the crater area. Then complete the stringer.
5. Pay very close attention to the electrode angles. Be sure you do not change angles during stopping and starting.
6. Watch your travel speed. It must be steady if the bead width is to remain constant.
7. Be sure the legs of the weld are of equal length.
8. Hold the same electrode angle with the second bead. Point the tip of the electrode so it is partially on the plate and partially on the first bead as shown in Figure 36-2. Slightly more than half of the filler metal should be on the stringer bead.
9. For the third stringer, lower the electrode angle to between 10 and 20 degrees as in Figure 36-2.
10. Complete the joint using the bead sequence shown in Figure 36-3.
11. Complete the last unwelded joint using the same six-bead technique.

84 SHIELDED METAL ARC WELDING PROCESS

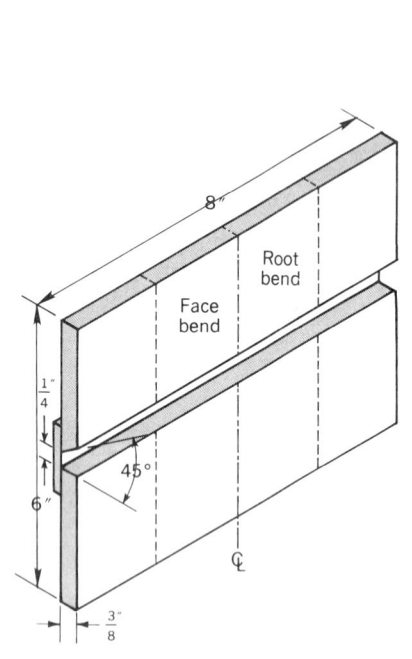

FIGURE 37-1

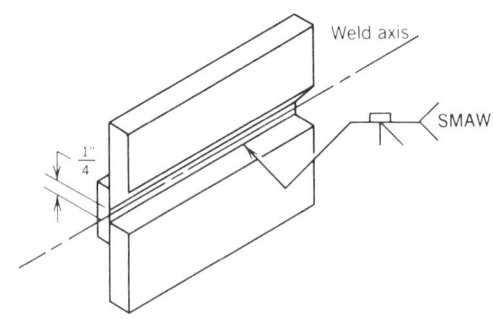

FIGURE 37-2

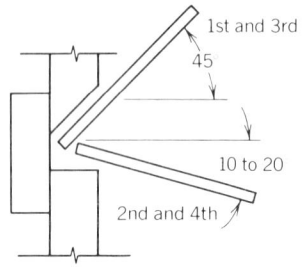

FIGURE 37-3

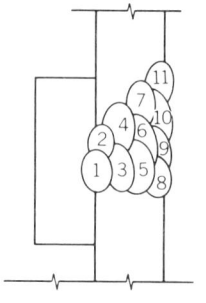

FIGURE 37-4

NAME	Single bevel butt	POSITION	Horizontal
ELECTRODE E-7018	DIAMETER ⅛"	AMPERAGE	115–165 AC or DCEP
MATERIAL	1 pc. ⅜" × 1½" × 8" carbon steel 2 pcs. ⅜" × 3" × 8" carbon steel		
PASSES Multiple	BEAD String	TIME (SEE INSTRUCTOR)	

Procedure 37

Single-Bevel Butt Joint, 2G Position with Backup Bar

OBJECTIVE

Upon completion of this lesson you should be able to weld a single-bevel butt joint with a backup bar in the horizontal position.

Text Reference: Section III, Lesson 5B; page 232.

PROCEDURE

1. Flame cut and grind a 45-degree bevel, on the long edge of one plate.
2. Remove all slag, mill scale, and so on from the area to be welded on all three plates.
3. Fit and tack the plates as shown in Figure 37-1. Tack at the ends of the joint.
4. Position the test piece as in Figure 37-2. The weldment may not be moved until all welding is completed.
5. Secure the weldment and workpiece clamp.
6. AC or DCEP. Adjust the welding current.
7. For the first pass hold an electrode work angle of about 45 degrees as shown in Figure 37-3. Use a drag angle of about 20 to 30 degrees. Do not let the puddle cover more than one-half of the root spacing. The electrode should penetrate deeply into the bottom plate and wash up on the backup bar as shown in Figure 37-4.
8. Clean the first bead thoroughly. Then run the second bead over the remaining root area and the beveled plate. Use an electrode work angle of about 10 to 20 degrees upward into the root as shown in Figure 37-3. For good fusion be sure to burn into the top plate, the remaining root space, and first bead as shown in Figure 37-4.
9. Clean the second bead thoroughly. Cover it with two stringers or one wide stringer depending on the width of the first bead. Stringers at the bottom plate must be made with an electrode angle of about 45 degrees if you are to get deep penetration. (Refer to Figures 37-3 and 37-4.)
10. All other stringers, until the last pass, should be at about a 10- to 20-degree upward angle. Follow the bead sequence shown in Figure 37-4. The exception to this is stringer 5 that requires the same angle as 1 and 3.
11. The next to last layer beads numbers 5, 6, and 7 in Figure 37-4 should be about 1/16 in. below the surface of the plate.
12. All the stringers in the finish pass beads numbers 8 to 11 in Figure 37-4 may be put in with an electrode angle of about 10- to 15-degree upward angle.

FIGURE 38-1

FIGURE 38-2

FIGURE 38-3

FIGURE 38-4

FIGURE 38-5

FIGURE 38-6

NAME **Open-root butt**	POSITION **Horizontal**
ELECTRODE: **E-6010** or **E-6011**; DIAMETER: **1/8″** / **1/8″**	AMPERAGE **75–125 AC or DCEP** **75–125 AC or DC**
MATERIAL **2 spacers 3/32″ × 3/4″ × 5″** **2 pcs. 3/8″ × 4″ × 8″ carbon steel**	
PASSES: **Multiple**; BEAD: **Weave**	TIME (SEE INSTRUCTOR)

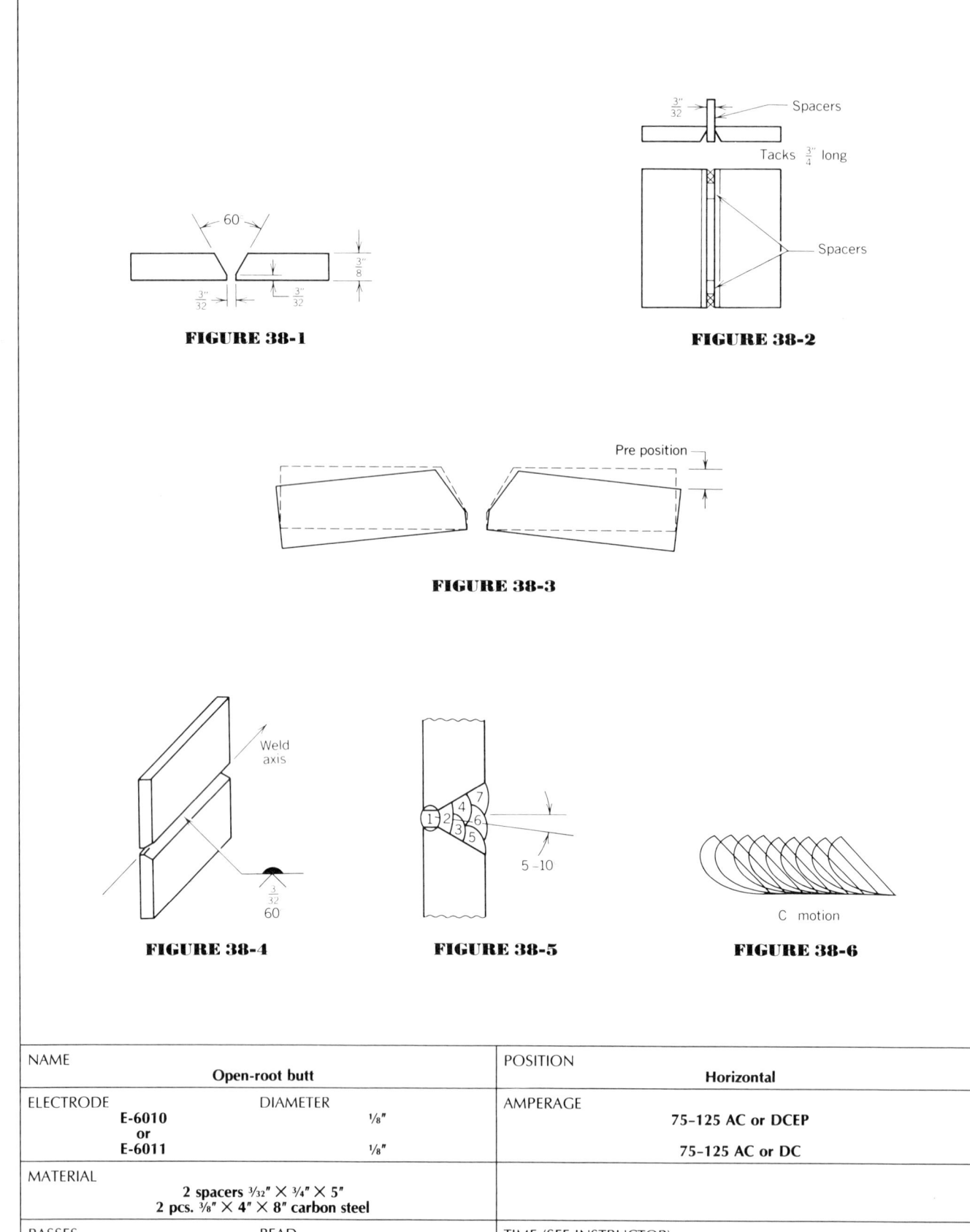

Procedure 38

Open-Root V-Groove, Butt Joint, 2G Position

OBJECTIVE

Upon completion of this lesson you should be able to weld open-root butt joints in the horizontal position.

Text Reference: Section III, Lesson 5B; page 233.

PROCEDURE

1. Use a hand cutting torch or mechanical flame cutter to bevel one of the long edges of each plate. Cut the bevel at a 30-degree angle. This makes a 60-degree included angle as in Figure 38-1.
2. Remove all slag, mill scale, and foreign matter by grinding or sanding.
3. Grind or file a 3/32 in. root face on each beveled edge as shown in Figure 38-1.
4. Lay the plates, face down, on a flat surface with the two spacers in place as in Figure 38-2 (top view).
5. Tack the two plates as indicated in Figure 38-2.
6. Remove the spacers and preposition the plates as in Figure 38-3.
7. Position the plate as in Figure 38-4; secure the plate and workpiece clamp.
8. AC or DCEP. Adjust the welding current.
9. After setting your heat on scrap metal, start the weld bead on the left. Use an electrode angle of about 5 to 10 degrees upward into the root. (See Figure 38-5.) Use a 10- to 15-degree drag angle.
10. Pause at the end of the tack until a keyhole appears.
11. When the keyhole appears, whip the electrode forward and down onto the wall of the lower bevel. Without stopping make a slight "C" motion and return to the keyhole. As you do this the puddle solidifies. The whipping allows the puddle to cool. It keeps the keyhole small and stops the puddle from spilling out the back of the joint.
12. With a little practice you will master this technique. If you pause too long in the puddle, you will "blow through" or spill the molten metal out of the rear of the joint. If you move too fast, you will not penetrate the joint properly.
13. The second pass is a stringer bead. Make it with the "C" motion. Take care to pause at the upper left hand portion of the "C" shown in Figure 38-6.
14. Finish the remainder of the joint with stringers as shown in Figure 38-5. Do not forget to leave about a 1/16 in. space between the face of beads number 3 and 4, the next to last pass, and the surface of the plate.

88 SHIELDED METAL ARC WELDING PROCESS

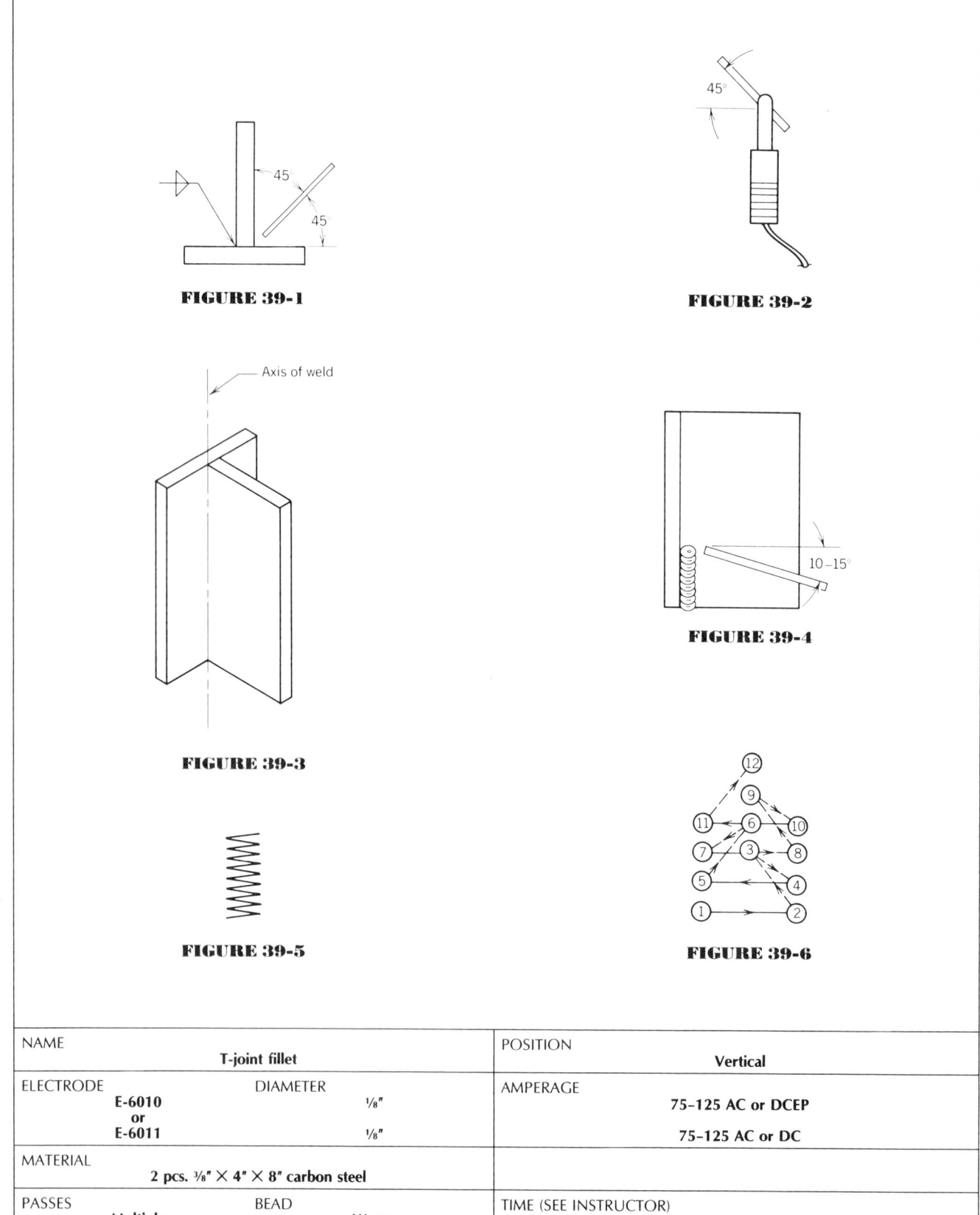

FIGURE 39-1

FIGURE 39-2

Axis of weld

FIGURE 39-3

FIGURE 39-4

FIGURE 39-5

FIGURE 39-6

NAME T-joint fillet	POSITION Vertical

ELECTRODE   DIAMETER
   E-6010   1/8"
   or
   E-6011   1/8"
	AMPERAGE
	75-125 AC or DCEP
	75-125 AC or DC

MATERIAL 2 pcs. 3/8" × 4" × 8" carbon steel	
PASSES BEAD Multiple Weave	TIME (SEE INSTRUCTOR)

Procedure 39

T-Joint Fillet, 3F Position, Weave Beads, Uphill (E-6010 or E-6011 Electrodes)

OBJECTIVE

Upon completion of this lesson you should be able to weld vertical fillets.

Text Reference: Section III, Lesson 5C; page 235.

PROCEDURE

1. Prepare, clean, and tack the plates to form a T-joint as in previous lessons. (See Figure 39-1.)
2. Place the electrode in the holder at a 45-degree angle as shown in Figure 39-2. Use this electrode position for all vertical welds.
3. Set your welding current on a piece of scrap metal in the vertical position.
4. Position the joint so the weld axis is vertical as in Figure 39-3.
5. Adjust the welding current.
6. Position yourself so the electrode is centered between the two plates as in Figure 39-1. Point the electrode slightly upward, at a 10 to 15 degree angle as in Figure 39-4.
7. First pass: Strike the arc at the bottom of the joint where the two plates meet. Hold the arc until the puddle is approximately ¼ to 5⁄16 in. wide where it joins the two plates. Whip the electrode up and to the right as in Figure 39-5. Hold a long arc as you move out of the puddle and as you drop back into it.
8. As the electrode drops back into the puddle, hold a normal length arc. Pause slightly and whip out of the puddle again.
9. Second pass: Start at the left side of the weld groove at point 1. (See Figure 39-6.) Pause slightly to allow the metal to form a puddle. When the puddle forms, move the electrode slowly to point 2 on the right side. Pause again to allow any undercut to fill in. Then whip the electrode to point 3 slightly above the center of the puddle. As the weld puddle changes color, return the electrode to point 4. Pause again while the undercut fills in, then slowly move across to point 5. Repeat the motions on the left as shown by points 6, 7, and 8. As you continue the sequence, remember the electrode is always returned to the side where it left the puddle.

Note: Always hold a long arc on both the upstroke and the down stroke so you do not deposit any metal during the electrode motion.

10. If the center of the weld is too high or if it sags, you are moving too slowly across the face of the weld. Move faster.
11. If the center of the weld is too low or has holes, you are moving too fast across the face. Move slower.
12. Move with a rhythm. Always pause at the sides of the groove.
13. As the electrode moves from side to side, make sure you stop with the electrode coating at the edge of the previous pass. The puddle will flow out past the edge and provide the correct pass width.
14. Finish the second pass in this manner; repeat the sequences for the third pass. The finish of the third pass should not exceed five electrode diameters in width.
15. Inspect for unevenness, undercut, holes, and bead contour.

90 SHIELDED METAL ARC WELDING PROCESS

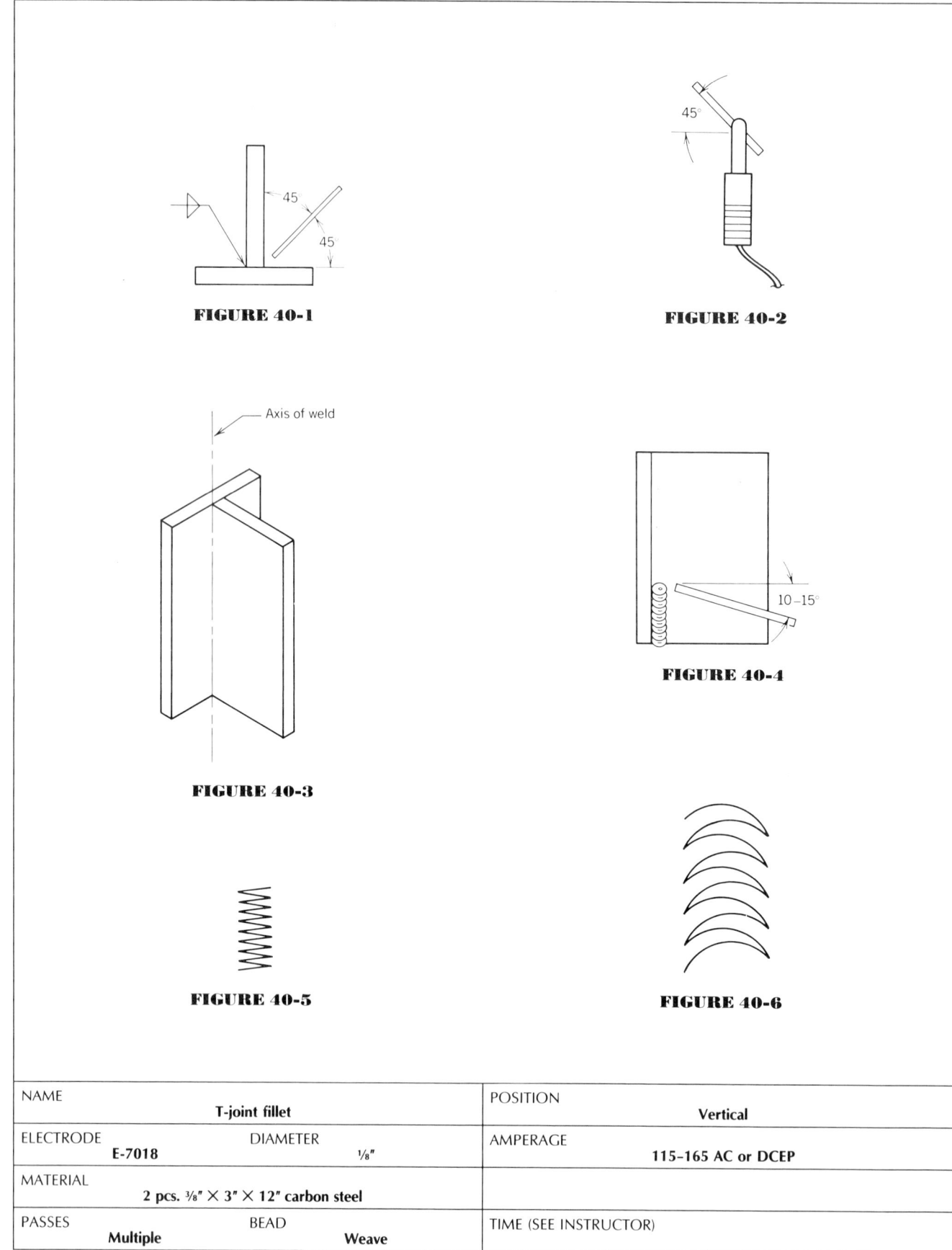

FIGURE 40-1

FIGURE 40-2

Axis of weld

FIGURE 40-3

FIGURE 40-4

FIGURE 40-5

FIGURE 40-6

NAME	T-joint fillet	POSITION	Vertical
ELECTRODE E-7018	DIAMETER ⅛"	AMPERAGE	115–165 AC or DCEP
MATERIAL	2 pcs. ⅜" × 3" × 12" carbon steel		
PASSES Multiple	BEAD Weave	TIME (SEE INSTRUCTOR)	

Procedure 40

T-Joint Fillet, 3F Position, Weave Beads, Uphill (E-7018 Electrode)

OBJECTIVE

Upon completion of this lesson you should be able to weld vertical fillets using low hydrogen electrodes.

Text Reference: Section III, Lesson 5C; page 236.

PROCEDURE

1. Prepare, clean, and tack plates to form a T-joint as shown in Figure 40-1.
2. Place the electrode in the holder at a 45-degree angle as shown in Figure 40-2.
3. Position the joint so the weld axis is vertical as in Figure 40-3.
4. Adjust the welding current.
5. Hold the electrode midway between the two plates as in Figure 40-1. Point the electrode slightly upward, at about a 0- to 5-degree angle, as in Figure 40-4.
6. First pass: Begin at the bottom of the joint. Start the arc with a tapping motion. This breaks the coating covering the end of the electrode.
7. Weave the electrode from side to side, advancing upward slightly to keep the bead progressing. (Refer to Figure 40-5.) The bead should only be about $5/16$ in. wide.
8. Pause at each side and watch for the weld puddle to sag in the center. If sag is present:
   a. Reduce the amperes,
   b. Increase the travel speed, or
   c. Hold a closer arc
9. Second pass: Clean the first bead; then strike the arc on the bottom center of the groove. Move quickly to the lower left edge of the first pass.
10. Pause until the puddle forms. Then move the electrode steadily to the right, drawing the puddle along.
11. Pause on the right when the edge of the electrode reaches the edge of the weld. The puddle will flow out another $3/32$ in. to $1/8$ in.
12. Use either the basic weave or the alternate weave shown in Figure 40-6.
13. Third pass: Weave another bead over the second pass. Cover slightly more than one-half of the right side of the weld. Watch for undercut on the right side of the weld.
14. Weave the fourth bead on the left side. Cover the plate from the second pass and overlap about $1/4$ in. of the plate from the third pass. Be careful of undercut, both on the left side and where this bead joins the other bead. Do not weave wider than three times the electrode diameter.

92 SHIELDED METAL ARC WELDING PROCESS

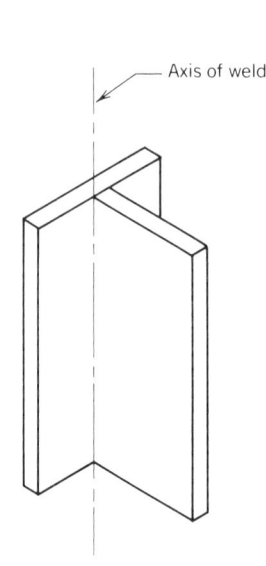

FIGURE 41-1

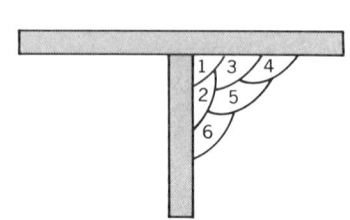

FIGURE 41-2

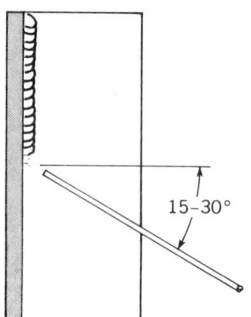

FIGURE 41-3

NAME	T-joint fillet	POSITION	Vertical down
ELECTRODE DIAMETER	E-6010 1/8″ or E-6011 1/8″	AMPERAGE	75–125 DCEP 75–125 AC or DC
MATERIAL	2 pcs. 3/8″ × 4″ × 8″ carbon steel		
PASSES BEAD	Multiple String	TIME (SEE INSTRUCTOR)	

Procedure 41

T-Joint Fillet, 3F Position, Stringer Beads, Downhill

OBJECTIVE

Upon completion of this lesson you should be able to weld vertical down fillets.

Text Reference: Section III, Lesson 5C; page 237.

PROCEDURE

1. Position the weldment as shown in Figure 41-1.
2. Adjust the welding current.
3. First pass: Point the electrode directly into the center of the joint. Use an upward angle of about 15 to 30 degrees as shown in Figure 41-2.
4. Strike the arc and weave down until you pass over the tack and until the end of the tack is reached. Hold a fairly long arc.
5. When you pass the tack, gently push the electrode into the root. The electrode should touch the workpiece.
6. Slowly drag the electrode down the joint making sure that the electrode angle remains the same.
7. If the slag begins to run ahead of the puddle, you must either increase your electrode angle and use the force of the arc to hold back the slag or increase your travel speed.
8. Run additional stringer beads using the bead sequence shown in Figure 41-3. Clean thoroughly between passes.
9. Complete the joint. Clean and examine it, then weld the second side using the same technique and bead sequence you used on the first side.

94 SHIELDED METAL ARC WELDING PROCESS

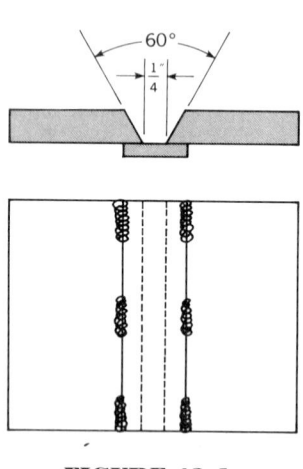

FIGURE 42-1

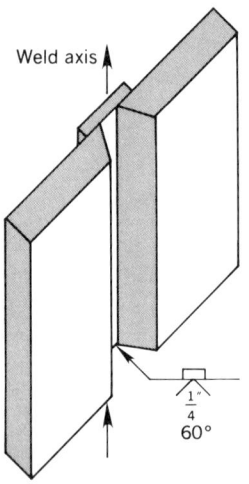

FIGURE 42-2

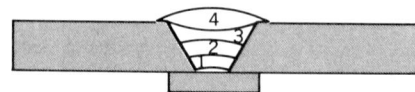

FIGURE 42-3

NAME	V-groove butt	POSITION	Vertical up
ELECTRODE E-7018	DIAMETER 1/8" or 5/32"	AMPERAGE	90–150 DCEP 110–230 DCEP
MATERIAL	1 pc. 1/4" × 1" × 12" carbon steel 2 pcs. 1/2" × 3" × 12" carbon steel		
PASSES Multiple	BEAD Weave	TIME (SEE INSTRUCTOR)	

Procedure 42

V-Groove Butt Joint, 3G Position with Backup Bar, Uphill

OBJECTIVE

Upon completion of this lesson you should be able to weld vertical v-groove joints using a backup bar.

Text Reference: Section III, Lesson 5C; page 237.

PROCEDURE

1. Put a 30-degree bevel on one 12 in. side of each plate.
2. Tack the assembly as shown in Figure 42-1.
3. Position the joint as shown in Figure 42-2.
4. Adjust the welding current.
5. Using a ⅛ in. electrode for the root pass begin at the bottom of the joint and establish a shelf on which to begin the vertical up bead.
6. Begin the root bead using a slight weave. Hesitate on the sides of the groove to make sure the weld does not undercut the sides of the groove.
7. Using the bead sequence shown in Figure 42-3 put in the first two layers with ⅛ in. electrodes, put in the third layer with 5/32 in. electrodes. This should bring the weld just below flush, with the surface of the plate. Complete the reinforcement with the ⅛ in. electrode being especially careful on the sides to hesitate enough to fill the joint and eliminate undercut.
8. Cool and clean the plate. Examine it for uniform width, even ripples, and absence of undercut along the sides of the weld.

96 SHIELDED METAL ARC WELDING PROCESS

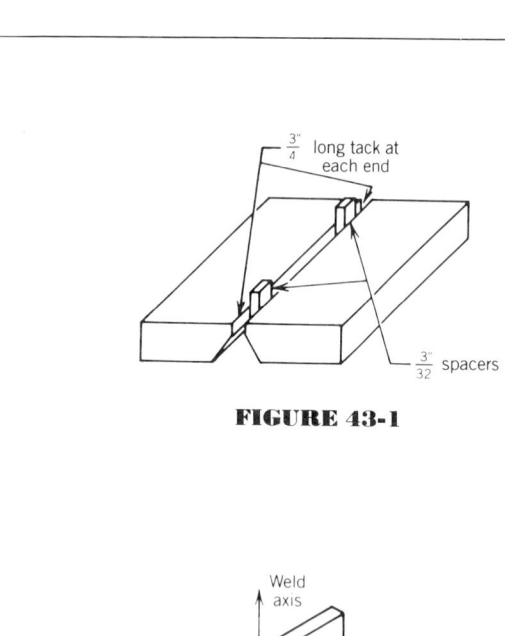

FIGURE 43-1

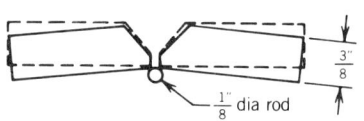

FIGURE 43-2

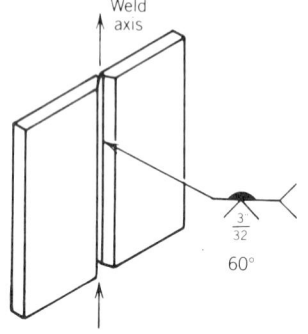

FIGURE 43-3

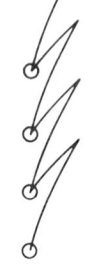

FIGURE 43-4

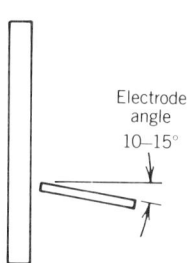

FIGURE 43-5

FIGURE 43-6

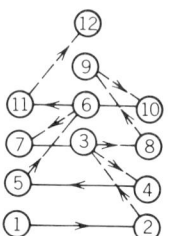

FIGURE 43-7

NAME	V-groove butt	POSITION	Vertical up
ELECTRODE E-6010 or E-6011	DIAMETER 1/8" or 1/8"	AMPERAGE	75–125 DCEP 75–125 AC or DCEP
MATERIAL	2 spacers 3/32" 2 pcs. 3/8" × 5" × 12" carbon steel		
PASSES Multiple	BEAD String and weave	TIME (SEE INSTRUCTOR)	

Procedure 43

Open-Root V-Groove Butt Joint, 3G Position, Stringer and Weave Beads

OBJECTIVE

Upon completion of this lesson you should be able to weld open-root vertical-butt joints.

Text Reference: Section III, Lesson 5C; page 238.

PROCEDURE

1. Bevel the long side of the plates with a 30-degree angle. This provides a final included angle of 60 degrees.
2. Remove the scale on the intended weld areas by grinding or sanding.
3. File a 3/32 in. root face land or shoulder on each bevel.
4. Fit it up with a 3/32 in. root gap and tack at each end as shown in Figure 43-1.
5. Preposition the plates slightly as in Figure 43-2 so you can fit a 1/8 in. electrode beneath the joint.
6. Position the joint at a comfortable height in the position shown in Figure 43-3.
7. Adjust the welding current.
8. Check your welding condition on scrap so that you can easily whip the electrode uphill as in Figure 43-4.
9. Start the arc on the bottom tack. Hold the arc at the beginning of the root opening until a keyhole appears. Use an electrode angle 10 to 15 degrees as shown in Figure 43-5.
10. First pass: Move the electrode upward in a slight whipping motion then return to the keyhole. Pause there until the keyhole fills and another opens above it. Continue to whip uphill.
11. If the keyhole does not appear, you are:
    a. Moving upward too rapidly
    b. Do not have sufficient arc current to penetrate
    c. Do not have the proper root opening
12. If the keyhole is too large, you are:
    a. Moving upward too slowly
    b. Have the current set too high
    c. Holding too long an arc
13. Clean the plate and run the second pass. Start at the bottom of the joint. Hold a normal arc and weld uphill using either the weave shown in Figure 43-6, or the triangle motion shown in Figure 43-7. Pause at the sides to eliminate undercut.
14. The next to last pass should be 1/32 to 1/16 in. below the plate surface. This gives you a guide for the last pass.
15. The final pass must be about 1/16 in. above the surface of the plate. Excess reinforcement can create a stress riser or notch that can lead to weld failure.

98 SHIELDED METAL ARC WELDING PROCESS

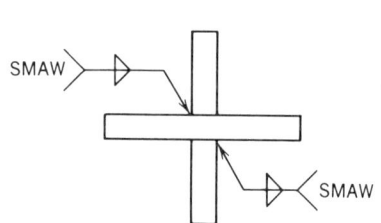

FIGURE 44-1

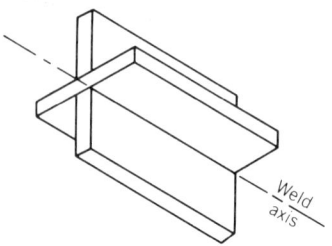

FIGURE 44-2

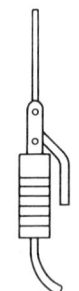

FIGURE 44-3

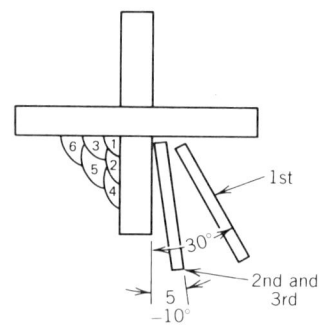

FIGURE 44-4

NAME	Fillet	POSITION	Overhead
ELECTRODE E-6010 or E-6011	DIAMETER 1/8" or 1/8"	AMPERAGE	75-125 DCEP 75-125 AC or DCEP
MATERIAL	1 pc. 3/8" × 4" × 12" carbon steel 2 pcs. 3/8" × 1 3/4" × 12" carbon steel		
PASSES Multiple	BEAD String	TIME (SEE INSTRUCTOR)	

Procedure 44

T-Joint Fillet, 4F Position, Stringer Beads (E-6010 or E-6011 Electrodes)

OBJECTIVE

Upon completion of this lesson you should be able to weld fillets in the overhead position.

Text Reference: Section III, Lesson 5D; page 240.

PROCEDURE

1. Wire brush the intended weld area of the plate.
2. Tack the plates together as shown in Figure 44-1.
3. Position the practice piece in the fixture as in Figure 44-2. Be sure the workpiece and workpiece clamp are secure.
4. Adjust the welding current.
5. Place the electrode in the holder as shown in Figure 44-3. Use a 30-degree work angle off the vertical plate as shown in Figure 44-4. Work with a drag angle of approximately 10 degrees.
6. Strike the arc and hold a fairly long arc. Begin a stringer and weld over the tack using the "C" motion. When the end of the tack is reached, shorten the arc. Then complete the stringer pass using the "C" motion. Pause at the upper left of the "C" to fill in the undercut. Direct the arc at the overhead plate as shown in Figure 44-4.
7. Clean the plate after the first pass thoroughly. Use a 5- to 10-degree work angle as in Figure 44-4. Run the next stringer so it covers about ⅛ in. of the vertical plate and about two-thirds of the first stringer.
8. Clean the plate and run the third stringer. Use a 5- to 10-degree work angle, as shown in Figure 44-4, favoring the top plate. About one-half to two-thirds of bead 2 should be seen. The legs of the weld should be equal with the face of the weld at a 45-degree angle to the plate surface.
9. Cool and clean the joint. If it is satisfactory, weld a third layer. The third layer should have three beads as shown in Figure 44-4. Clean the joint thoroughly after each bead.

100 SHIELDED METAL ARC WELDING PROCESS

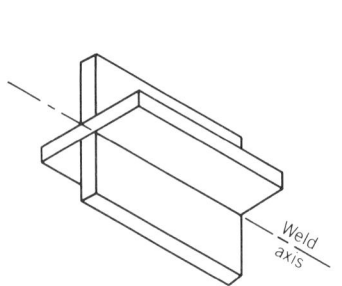

FIGURE 45-1

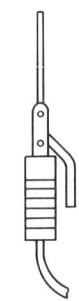

FIGURE 45-2

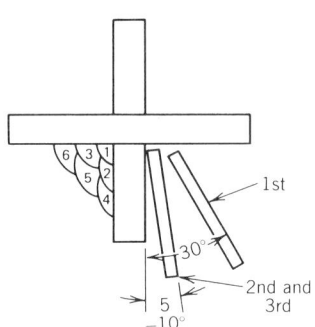

FIGURE 45-3

NAME	Fillet	POSITION	Overhead
ELECTRODE E-7018	DIAMETER ⅛"	AMPERAGE	115–165 AC or DCEP
MATERIAL	Use the two joints from Procedure 44		
PASSES Multiple	BEAD String	TIME (SEE INSTRUCTOR)	

Procedure 45

T-Joint Fillet, 4F Position, Stringer Beads (E-7018 or E-7024 Electrodes)

OBJECTIVE

Upon completion of this lesson you should be able to weld overhead fillets using low hydrogen electrodes.

Text Reference: Section III, Lesson 5D; page 241.

PROCEDURE

1. Clamp the practice plates in position as shown in Figure 45-1. Be sure the workpiece and workpiece clamp are secure.
2. Adjust the welding current.
3. Place the electrode in the holder as shown in Figure 45-2. Strike an arc using the tapping method. Start the first bead at the left side of the joint with the electrode at a 30-degree work angle off the vertical plate as in Figure 45-3. Use a drag angle of approximately 10 degrees.
4. Hold a slightly long arc and weld over the tack using the "C" motion. Make sure that you do not oscillate excessively or leave the puddle.
5. When the end of the tack is reached, hold as short an arc as possible, without stubbing out, and complete the stringer. Move the electrode along, pulling steadily, using the "C" motion. But never allow the tip of the electrode to leave the puddle. "Play" the arc on the overhead plate by pausing slightly at the top left of the "C" motion.
6. Clean thoroughly. Use the slag hammer to remove the slag, then brush the joint thoroughly. After brushing inspect carefully for any slag you might have missed. The slag left by low hydrogen electrodes is sometimes difficult to remove, especially if there is an undercut area where it can adhere.
7. Run the second stringer, using the same electrode angles as with the first pass. Cover about two-thirds of the first pass and about ⅛ in. of the vertical plate as shown in Figure 45-3.
8. Clean thoroughly and run the third stringer. Use a work angle of 5 to 10 degrees off the vertical plate as shown in Figure 45-3. About one-half to two-thirds of the second stringer should be uncovered. The weld legs should be equal. The face of the weld should make a 45-degree angle with the plate surfaces.
9. Run a third layer, consisting of three stringers as shown in Figure 45-3.
10. Complete the T-bar assembly for more practice.

102 SHIELDED METAL ARC WELDING PROCESS

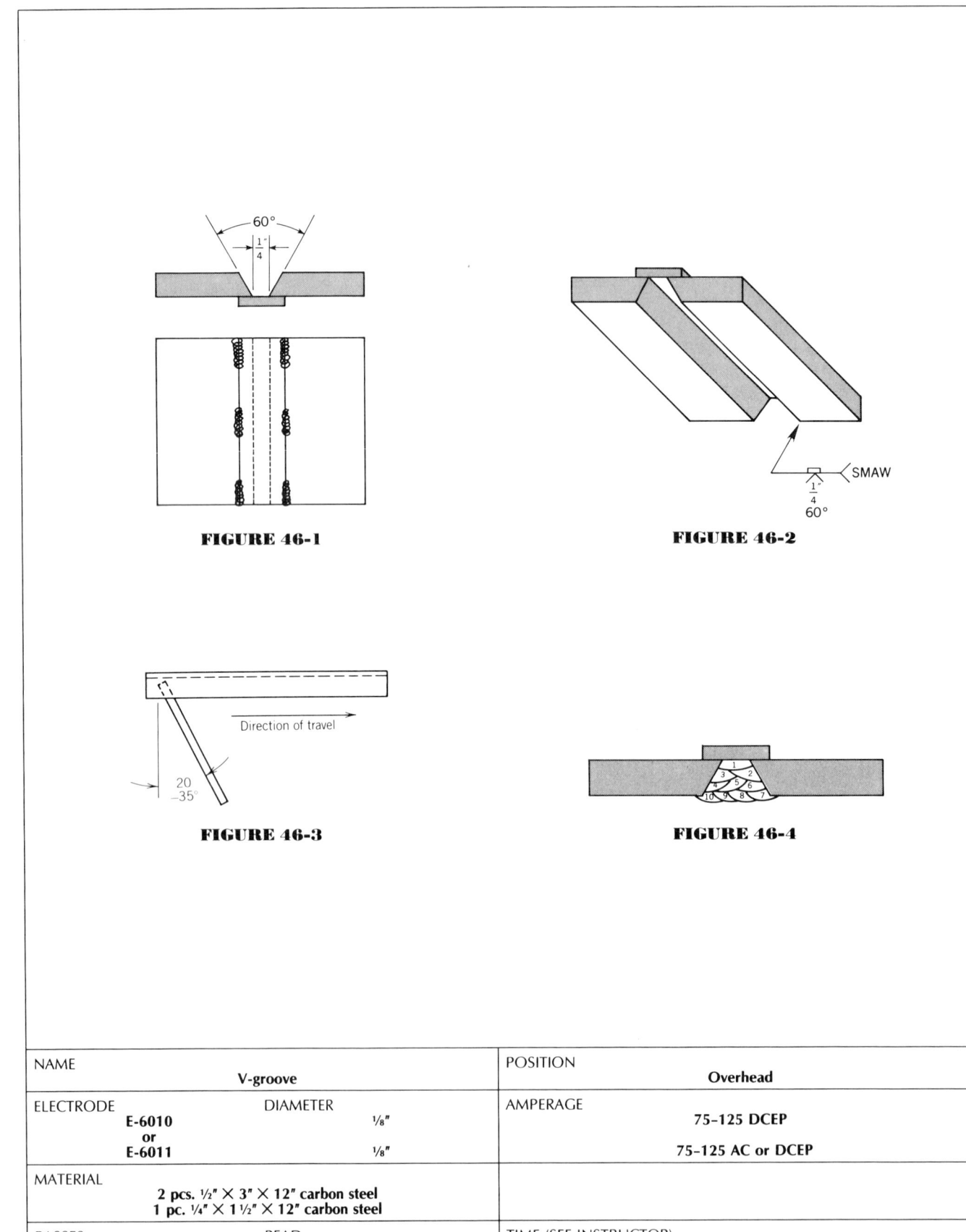

FIGURE 46-1

FIGURE 46-2

FIGURE 46-3

FIGURE 46-4

NAME	V-groove	POSITION	Overhead
ELECTRODE DIAMETER	E-6010 1/8" or E-6011 1/8"	AMPERAGE	75-125 DCEP 75-125 AC or DCEP
MATERIAL	2 pcs. 1/2" × 3" × 12" carbon steel 1 pc. 1/4" × 1 1/2" × 12" carbon steel		
PASSES BEAD	Multiple String	TIME (SEE INSTRUCTOR)	

Procedure 46

V-Groove Butt Joint, 4G Position with Backup Bar (E-6010 or E-6011 Electrodes)

OBJECTIVE

Upon completion of this lesson you should be able to weld overhead grooves with a backup bar.

Text Reference: Section III, Lesson 5D; page 241.

PROCEDURE

1. Put a 30-degree bevel on one 12 in. side and tack the plates together as shown in Figure 46-1. Position them as shown in Figure 46-2.
2. Adjust the welding current.
3. Place the electrode in the holder as in Procedure 45. Start at the left side of the joint. Hold the electrode perpendicular to the plate. Use a drag angle of approximately 20 to 35 degrees as shown in Figure 46-3.
4. Advance the electrode along the joint in a straight line using a slight weave to ensure both sides of the joint are welded. Keep the arc short and travel speed even.
5. Clean the bead thoroughly. Pay attention to the toe of the weld. All slag must be removed from this area.
6. Complete the remainder of the joint with stringer beads using a bead sequence as shown in Figure 46-4. Pay particular attention that the toe of the weld fuses into the other beads and base metal. Make sure the final layer finishes up with a maximum reinforcement of $3/32$ in.

104 SHIELDED METAL ARC WELDING PROCESS

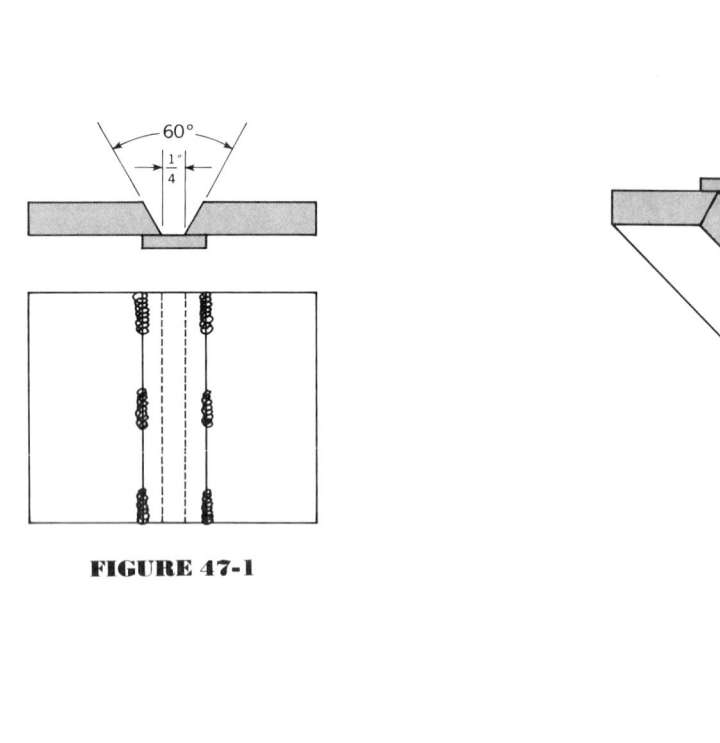

FIGURE 47-1

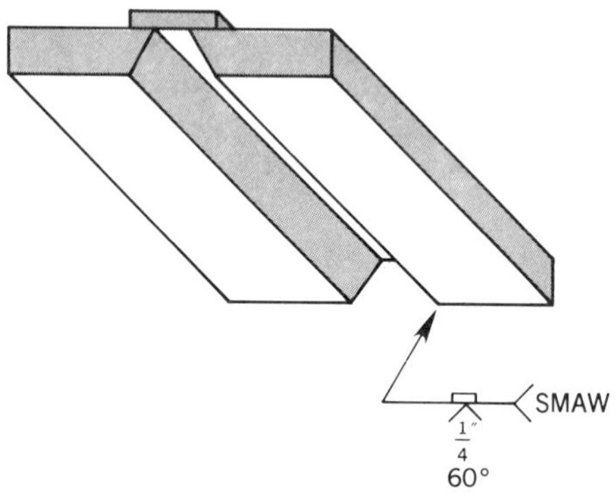

FIGURE 47-2

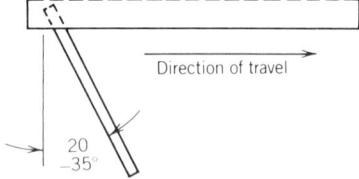

FIGURE 47-3

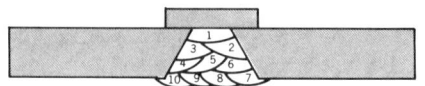

FIGURE 47-4

NAME			POSITION	
	V-groove			Overhead
ELECTRODE		DIAMETER	AMPERAGE	
E-7018		⅛"		115–165 DCEP
MATERIAL	2 pcs. ½" × 3" × 12" carbon steel 1 pc. ¼" × 1½" × 12" carbon steel			
PASSES		BEAD	TIME (SEE INSTRUCTOR)	
Multiple		String		

Procedure 47

V-Groove Butt Joint, 4G Position with Backup Bar (E-7018 Electrode)

OBJECTIVE

Upon completion of this lesson you should be able to weld overhead grooves using low hydrogen electrodes.

Text Reference: Section III, Lesson 5D; page 242.

PROCEDURE

1. Put a 30-degree bevel on one side and tack the plates together as shown in Figure 47-1. Position them as shown in Figure 47-2.
2. Adjust the welding current.
3. Start at the left side of the joint. Hold the electrode perpendicular to the plate using a drag angle of 20 to 35 degrees. (See Figure 47-3.)
4. Advance the electrode along the joint in a straight line at an even travel speed, using a slight weave ensuring both sides of the joint are properly fused. Do not use a whipping motion when using low hydrogen electrodes. Doing so will cause porosity.
5. Clean each bead thoroughly being careful to ensure the toe of the weld is thoroughly cleaned and free of slag. Any remaining slag will become trapped in subsequent beads.

Note: The bead appearance of the low hydrogen electrode is considerably smoother and more even than the surface obtained with E-6010 or E-6011 electrodes.

6. Complete the joint using stringer beads with a bead sequence as shown in Figure 47-4. Keep the second to last layer (beads 4, 5, and 6) slightly below flush so the reinforcement will not exceed 3/32 in. when finished.

106 SHIELDED METAL ARC WELDING PROCESS

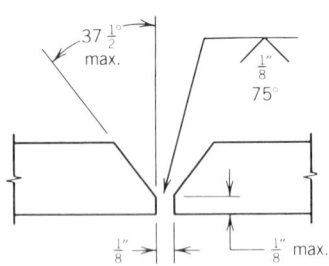

FIGURE 48-1

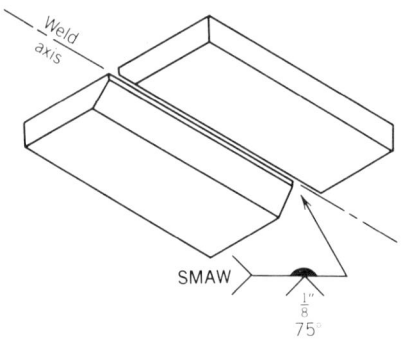

FIGURE 48-2

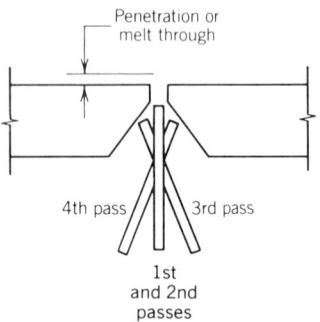

FIGURE 48-3

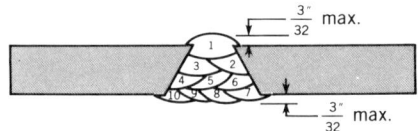

FIGURE 48-4

NAME	V-groove	POSITION	Overhead
ELECTRODE E-6010 E-7018	DIAMETER 1/8" 1/8"	AMPERAGE	75-125 DCEP 115-165 DCEP
MATERIAL	2 pcs. ½" × 5" × 10" carbon steel		
PASSES Multiple	BEAD String	TIME (SEE INSTRUCTOR)	

Procedure 48

Open-Root V-Groove Butt Joint, 4G Position

OBJECTIVE

Upon completion of this lesson you should be able to weld overhead open-root groove welds using low hydrogen electrodes.

Text Reference: Section III, Lesson 5D; page 242.

PROCEDURE

1. Be sure that the bevel angle and root face are prepared according to Figure 48-1. Clean all intended weld areas and tack the plates as shown in Figure 48-1.
2. Position the test piece as shown in Figure 48-2. Be sure the test piece and workpiece clamp are secure. Adjust the welding current.
3. Use the E-6010 electrode and a drag angle of approximately 10 degrees to whip in the first pass. Make sure that you open a keyhole. Keep it open along the entire length of the joint.
4. Hold a very close arc. Push the electrode slightly through the keyhole. This will force the molten metal to pile up on the top side of the joint (approximately $1/16$ to $3/32$ in. in height).
5. Clean the first pass thoroughly. Run a stringer by drawing the electrode along the joint for the second pass. Use E-7018 electrodes for this and the rest of the passes. Make sure the arc is running fairly "hot." The heat is needed to burn out any wagon tracks or deep grooves at the weld toes. Stay in the puddle.
6. The third layer will be made with two stringers.

Remember: Favor the toe area of the weld. Point the electrode slightly in the direction you want the metal to flow toward, (See Figure 48-3.)

7. Complete the remainder of the joint using the stringer method. Be careful of producing undercut or excessive reinforcement on the finish pass. See Figure 48-4 for the bead sequences and amount of weld reinforcement.

Shielded Metal Arc Welding of Sheet Metal

Welding sheet metal, especially that below 16 gauge, is extremely difficult with the shielded metal arc welding process. Thin sheet metal is better suited for welding by the gas metal arc, the gas tungsten arc, and the oxyacetylene processes. Gas metal arc welding is a fast and economical process. Gas tungsten arc welding, although slower, has the ability to produce exceptionally sound welds in the "hard to weld" metals such as stainless steel, aluminum, magnesium, titanium, copper, and so on.

The oxyacetylene process is an old standby; however, it lacks the speed needed for most types of production welding. This method also increases the heat input and causes the metal to warp.

A skilled shielded metal arc welder can successfully weld some sheet metal gauges. However, 16 gauge is the thinnest that can be welded successfully without too much distortion or burn-through.

GENERAL INSTRUCTIONS

Joint preparation and fit-up are of great importance in sheet metal welding. Slight openings in fillet joints or butt-joint faces will cause the arc to burn away the sheet metal edges during the welding operation. It is important to observe the following instructions as you complete each procedure in this section:

1. Shear or saw all metal to be joined. Flame cutting is a poor choice. Even if the welder is skillful enough to eliminate the slag from the cutting operation, the metal will warp from the heat. A good fit-up is difficult, if not impossible, to obtain with warped sheet metal.
2. Tacks on sheet metal should be close together. Depending on the gauge of the metal and the type of joint, they should be placed no more than 1 to 1½ in. apart. Tacks that are farther apart will allow the joint to open from the heat making it difficult to weld the joint.
3. Know your electrodes; choose the proper one for the job.

MATERIALS AND EQUIPMENT

For this section the material and equipment will be as follows:

1. Welding shield
2. Safety glasses
3. Protective leathers and gloves
4. Chipping hammer and wire brush
5. Personal welding equipment
6. Pliers
7. E-6013 electrodes, ³⁄₃₂ in. diameter

110 SHIELDED METAL ARC WELDING PROCESS

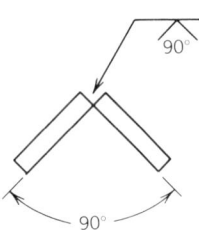

FIGURE 49-1

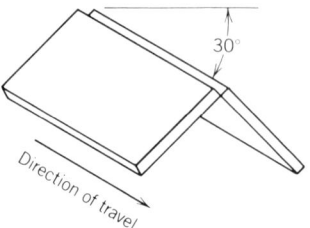

FIGURE 49-2

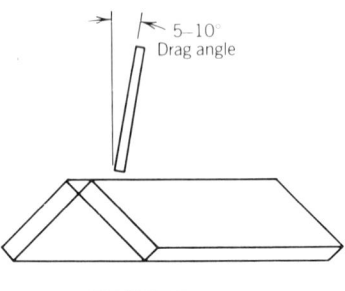

FIGURE 49-3

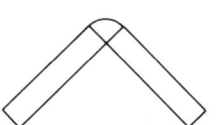

FIGURE 49-4

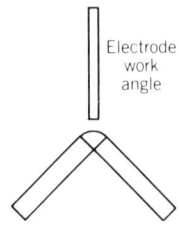

FIGURE 49-5

NAME	Corner joint	POSITION	Flat
ELECTRODE E-6013	DIAMETER 3/32"	AMPERAGE	45-90 DCEN
MATERIAL	2 pcs. 16 gauge × 2" × 10" carbon steel		
PASSES Single	BEAD String	TIME (SEE INSTRUCTOR)	

Procedure 49

Welding an Outside-Corner Joint in the Flat or Downhand Position

OBJECTIVE

Upon completion of this lesson you should be able to weld outside corner joints in sheet metal.

Text Reference: Section III, Lesson 6A; page 245.

PROCEDURE

1. Adjust the welding current on 16-gauge scrap sheet.
2. Tack the two pieces of sheet metal to form the open-corner joint shown in Figure 49-1. Tacks on sheet metal differ from those used on plate. Sheet metal tacks should be small; they are no larger than arc strikes. Apply them by striking and maintaining an arc for two or three counts and then break the arc.
3. After tacking, secure the weldment as shown in Figure 49-2.
4. Remove all slag from the tacks and wire brush the intended weld areas.
5. Place the electrode in the holder at a 90-degree angle.
6. Starting at the highest point of the joint, strike the arc and hold a normal arc gap. Move steadily along the joint with a perpendicular work angle as in Figure 49-3. Use a 5- to 10-degree drag angle as shown in Figure 49-4. The electrode can also be dragged along in contact with the metal. However, the electrode drag angle must be increased to approximately 25 to 30 degrees.
7. Stop midway along the joint. Remove the slag and inspect the bead. It should be slightly convex with no overlap at the toes of the weld. It should be free of surface holes and slag inclusion. Figure 49-5 shows the desired bead profile.
8. If the face of the bead is convex and there is overlap, there is too much filler metal. This happens when your rate of travel is too slow, your heat is set too high, or you have too much drag angle. If the bead is flat or concave, there is too little metal. This happens when your rate of travel is too fast or your amperage is too high. If the bead is high, just sitting on top of the joint, your amperage is too low. Change your welding conditions accordingly.
9. After your make the necessary adjustments, pick up the weld at the crater. Hold a long arc to start, then complete the joint. Inspect as before and continue to weld practice pieces until you can produce acceptable welds.

112 SHIELDED METAL ARC WELDING PROCESS

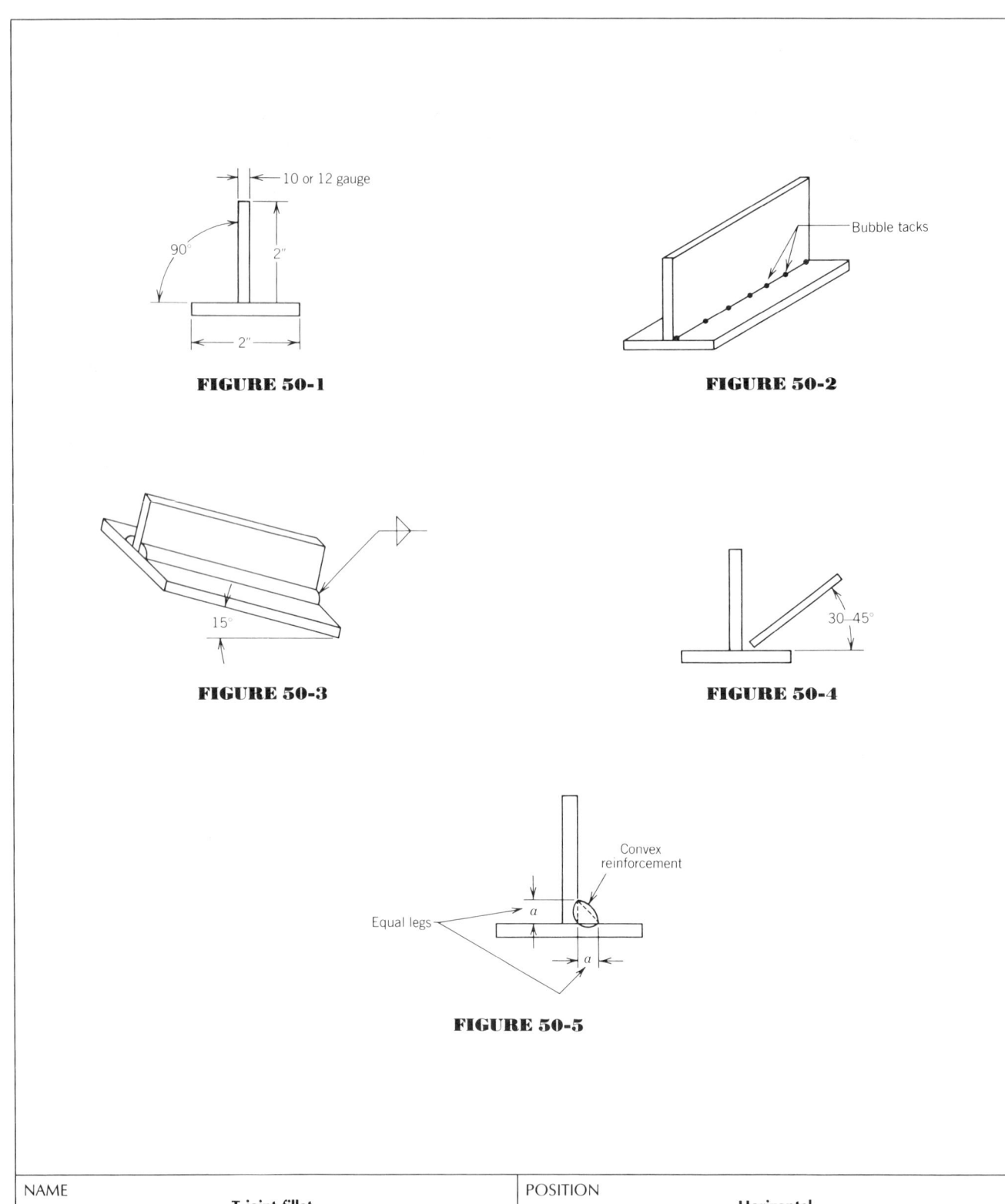

FIGURE 50-1

FIGURE 50-2

FIGURE 50-3

FIGURE 50-4

FIGURE 50-5

NAME	T-joint fillet	POSITION	Horizontal
ELECTRODE E-6012	DIAMETER 1/8"	AMPERAGE	80-140 AC or DCEN
MATERIAL	2 pcs. 10 or 12 gauge × 2" × 10" carbon steel		
PASSES Single	BEAD String	TIME (SEE INSTRUCTOR)	

Procedure 50

T-Joint Fillet, 2F Position

OBJECTIVE

Upon completion of this lesson you should be able to weld horizontal fillets on sheet metal using an E-6012 electrode.

Text Reference: Section III, Lesson 6B; page 246.

PROCEDURE

1. Adjust the welding current.
2. Adjust the current on 10- or 12-gauge sheet metal.
3. Clean the intended weld areas.
4. Fit-up the two pieces to form a T-joint as in Figure 50-1. Make sure the vertical piece is perpendicular.
5. Place one tack at each end as in Figure 50-2. Place smaller tacks between the end tacks, at about 1 in. intervals. These tacks are sometimes referred to as bubble tacks. They should be round and slightly larger in diameter than the electrode.
6. Secure the tacked pieces in position as shown in Figure 50-3. Tilt it, up to 15 degrees in the direction of travel. This will let you weld faster and reduce the chance of undercut.
7. Clean the slag left by the tacking procedure and start at the left side. Place the electrode in the holder, at a 45-degree angle as you did when welding horizontal fillets on plate. The electrode work angle should be between 30 and 45 degrees as shown in Figure 50-4. Use a 10- to 15-degree lead or drag angle.
8. Maintain a puddle slightly larger than the electrode diameter (maximum 3/16 in.), by holding a close arc with the electrode pointed into the heel of the joint. Favor the bottom piece slightly.
9. You can either hold a close arc or drag the electrode. If you drag the electrode, touch the leading edge of the coating to the surface of the joint. Watch the upper trailing edge of the puddle. If it does not fill in completely, you should travel slower. Otherwise, the vertical leg will be undercut.
10. When you reach the halfway point, stop. Clean the weld and remove the slag. Examine the bead to be sure you are fusing both pieces. The weld face should be at a 45-degree angle. The legs should be equal and the surface flat or slightly convex. (See Figure 50-5.)
11. Restart the arc using the same electrode, and complete the weld.
12. When you have completed the weld, cool and clean it thoroughly. Then weld the other side.

114 SHIELDED METAL ARC WELDING PROCESS

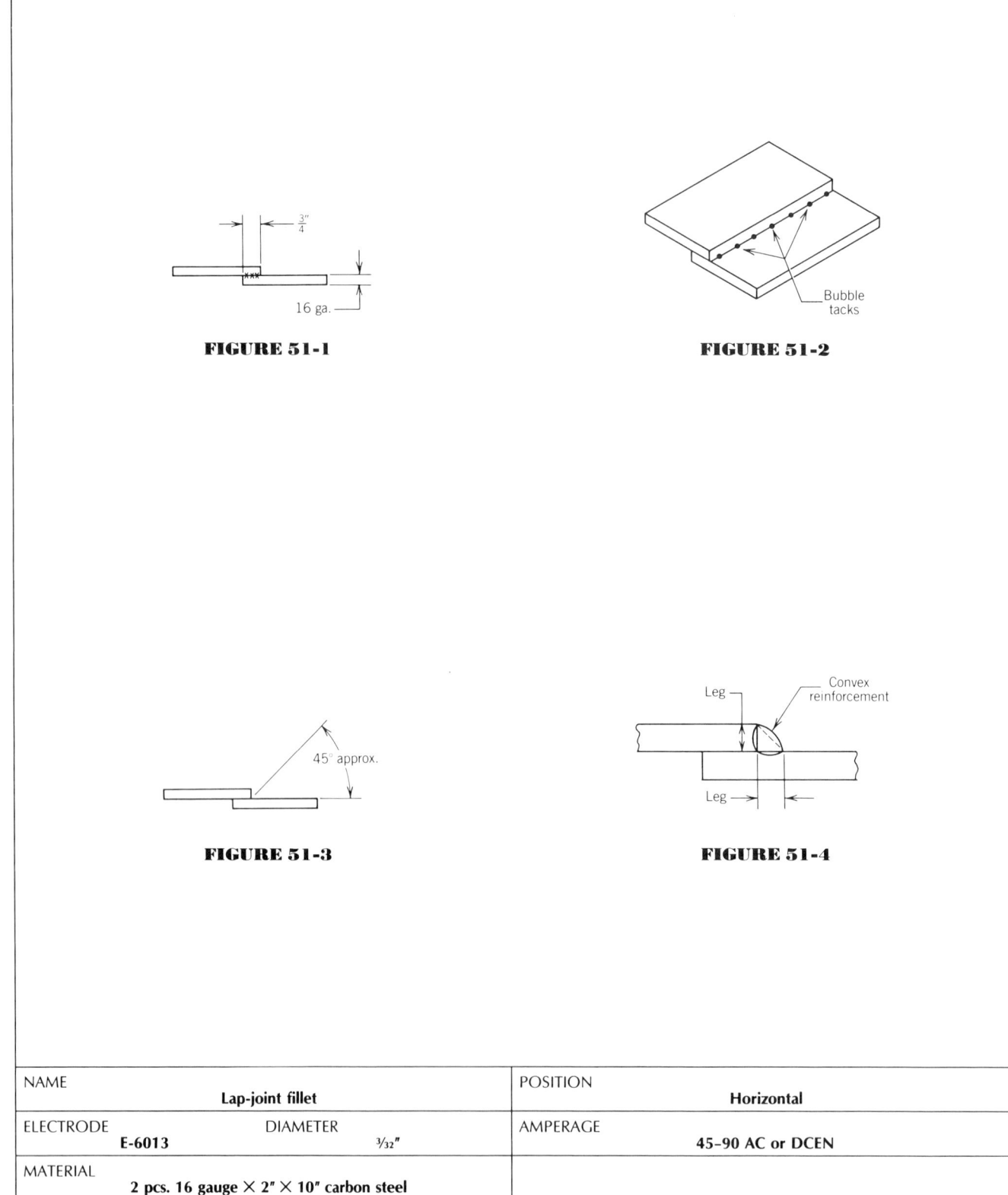

FIGURE 51-1

FIGURE 51-2

FIGURE 51-3

FIGURE 51-4

NAME Lap-joint fillet	POSITION Horizontal
ELECTRODE E-6013 DIAMETER 3/32"	AMPERAGE 45-90 AC or DCEN
MATERIAL 2 pcs. 16 gauge × 2" × 10" carbon steel	
PASSES Single BEAD String	TIME (SEE INSTRUCTOR)

Procedure 51

Lap Joint Fillet, 2F Position

OBJECTIVE

Upon completion of this lesson you should be able to weld sheet metal lap joints in the horizontal position.

Text Reference: Section III, Lesson 6B; page 247.

PROCEDURE

1. Adjust the welding current.
2. Clean all intended weld areas. Then tack the two pieces on the ends as shown in Figure 51-1.
3. Make sure there are no openings along the joint and place bubble tacks as in Figure 51-2. The closely spaced tacks will keep the joint from opening up as the weld progresses.
4. Secure the tacked pieces in position as in Figure 51-2. You may tilt the pieces slightly (a maximum of 15 degrees) toward the direction of travel. If you do, welding will proceed at a slightly faster pace and you will minimize the chance of undercut.
5. Remove the slag left by the tacking procedure and place the electrode in the holder at a 45-degree angle.
6. Start at the left side of the joint. The electrode work angle should be approximately 45 degrees as shown in Figure 51-3. Use about a 10- to 15-degree lead or drag angle. Maintain the puddle slightly larger than the legs of the joint.

Note: Do not burn away an excessive amount of the toe at the top of the vertical leg.

7. Pull the puddle, with a uniform width, along the entire length of the joint. With a $3/32$ in. diameter electrode it is very difficult to keep the weld small enough for metal of this thickness. If you have too much trouble, try using $1/16$ in. diameter electrodes.
8. Check the completed weld for surface holes, slag inclusion, and discontinuities, including spaces left unwelded. The weld should be completed in a single pass. It should have equal legs and a flat or slightly convex face as shown in Figure 51-4.
9. Cool and clean the sample. Then weld the opposite side.

116 SHIELDED METAL ARC WELDING PROCESS

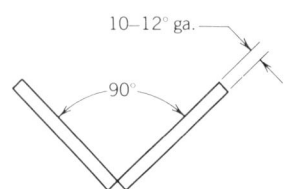

FIGURE 52-1

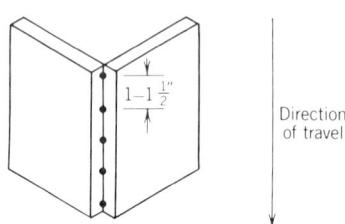

FIGURE 52-2

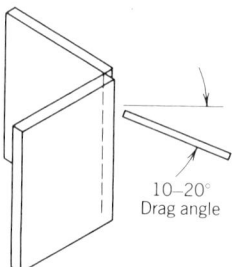

FIGURE 52-3

NAME		POSITION	
	Corner joint		Vertical
ELECTRODE	DIAMETER	AMPERAGE	
E-6011	3/32"		40–80 DCEN
MATERIAL			
2 pcs. 10 or 12 gauge × 2" × 10" carbon steel			
PASSES	BEAD	TIME (SEE INSTRUCTOR)	
Single	String		

Procedure 52

Outside-Corner Joint, 3G Position (E-6011 Electrode)

OBJECTIVE

Upon completion of this lesson you should be able to weld outside-corner joints in the vertical position.

Text Reference: Section III, Lesson 6C; page 248.

PROCEDURE

1. Adjust the welding current.
2. Tack weld the joint as in Figures 52-1 and 52-2. Place the tacks 1 to 1½ in. apart. Be sure there are enough tacks; otherwise the joint will open during welding.
3. Secure the weldment in the 3G vertical position as in Figure 52-2.
4. Remove all slag from the tacking procedure and clean the intended weld area. The welding will be from the top of the joint toward the bottom.
5. Place the electrode in the holder at a 45-degree angle.
6. Start at the top of the joint. Strike an arc and move steadily toward the bottom using about a 10- to 20-degree drag angle as shown in Figure 52-3. Point the electrode straight into the joint.
7. As you follow the groove it may be necessary to weave the electrode slightly from side to side. This will fuse the toes of the weld. If the puddle is wide enough, the weaving is not necessary.
8. Hold a fairly long arc. Do not be concerned by any flux that moves ahead of the puddle. Whatever flux is present will be light, and it will have no effect on the advancing weld.
9. Concentrate on completing the weld without stopping. If a pause is necessary, be sure to clean the crater area thoroughly before continuing.
10. The contour of the weld can range from a flat face to slightly convex face.
11. You can continue to practice this joint by adding pieces to the weldment. Be sure to clean all plate edges and intended weld areas before you begin. (See Figure 52-1.)

118 SHIELDED METAL ARC WELDING PROCESS

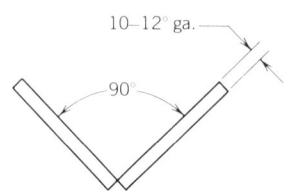

FIGURE 53-1

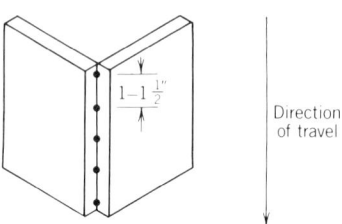

FIGURE 53-2

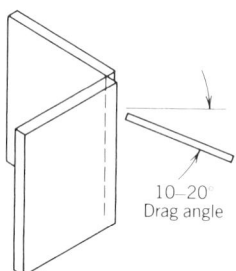

FIGURE 53-3

NAME			POSITION	
	Corner joint			Vertical
ELECTRODE		DIAMETER	AMPERAGE	
E-6013		⅛″		80–130 DCEN
MATERIAL				
2 pcs. 10 or 12 gauge × 2″ × 10″ carbon steel				
PASSES		BEAD	TIME (SEE INSTRUCTOR)	
Single		String		

Procedure 53

Outside-Corner Joint, 3G Position (E-6013 Electrode)

OBJECTIVES

Upon completion of this lesson you should be able to weld vertical corner joints using E-6013 electrodes.

Text Reference: Section III, Lesson 6C; page 249.

PROCEDURE

1. Adjust the welding current.
2. Clean all intended weld areas.
3. Fit-up and tack two pieces to form an outside-corner joint as in Figures 53-1 and 53-2. Make sure the pieces are at a right angle, as indicated. Place a tack at each end and enough tacks in between to keep the joint from opening during the welding procedure. Clean the slag left by the tacking procedure.
4. Secure the weldment in position as shown in Figure 53-2.
5. Place the electrode in the holder at a 45-degree angle.
6. Start to weld from the top. Aim the electrode straight into the joint. The drag angle should be between 10 and 25 degrees as shown in Figure 53-3.
7. Using the downhill technique drag the electrode downward using a slight weaving motion if necessary. Spread the weld pool to cover the entire joint.
8. Do not oscillate the electrode excessively. Excessive oscillation will cause the weld to be too wide. The weld width should be slightly more than the joint width.
9. If the slag runs ahead of the molten puddle, you should stop. Chip the weld at the crater, clean it, and start over.

Remember: Hold a long arc when restarting. Completely fill in the crater before continuing the bead.

10. Experiment a little with the length of the arc. Try a short length arc and a medium arc. Note the difference, then use the arc length you feel is the best for you. This holds true for electrode angles. Some angles improve the appearance of the weld and the ease with which it is welded.

120 SHIELDED METAL ARC WELDING PROCESS

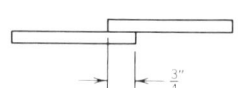

FIGURE 54-1

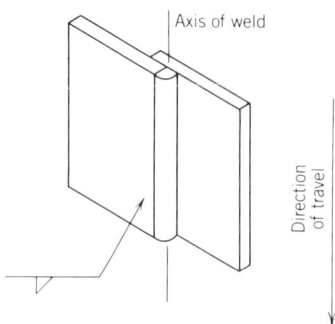

FIGURE 54-2

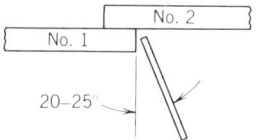

FIGURE 54-3

NAME	Lap joint	POSITION	Vertical
ELECTRODE E-6013	DIAMETER 3/32"	AMPERAGE	45-90 DCEN
MATERIAL	2 pcs. 16 gauge × 2" × 10" carbon steel		
PASSES Single	BEAD String	TIME (SEE INSTRUCTOR)	

Procedure 54

Lap Joint Fillet, 3F Position

OBJECTIVE

Upon completion of this lesson you should be able to weld vertical lap joints on sheet metal.

Text Reference: Section III, Lesson 6C; page 249.

PROCEDURE

1. Adjust the welding current.
2. Fit-up and tack the joint on the ends as in Figure 54-1. Place enough tacks between the ends so the joint will not open from the heat of the welding.
3. Remove all slag from the tacking procedure and clean all intended weld areas.
4. Secure the weldment as shown in Figure 54-2. The direction of welding will be downward, from the top of the joint.
5. Place the electrode in the holder at a 45-degree angle. Start at the top of the joint using a 10- to 20-degree drag angle, a work angle of 20 to 25 degrees, off the vertical edge of the overlapping sheet as shown in Figure 54-3.
6. Use the drag technique. Hold a fairly short arc. The puddle should be large enough in diameter to fuse the entire joint.
7. Favor the surface of sheet 2 (see Figure 54-3). Watch the left side of the puddle closely and make sure it fuses the leading edge of sheet 1. If you move too slowly or weave excessively, you will burn away the leading edge of sheet 1. This is not acceptable.
8. It is very difficult to obtain a finished weld with equal legs on thin sheet metal while using a 3/32-in. diameter electrode. A better electrode choice would be 1/16 in. diameter. If 1/16 in. electrodes are available, use them to practice this joint.
9. Usually the leg of the fillet on sheet 2 is slightly longer. This is acceptable as long as the other leg is fused properly.

122 SHIELDED METAL ARC WELDING PROCESS

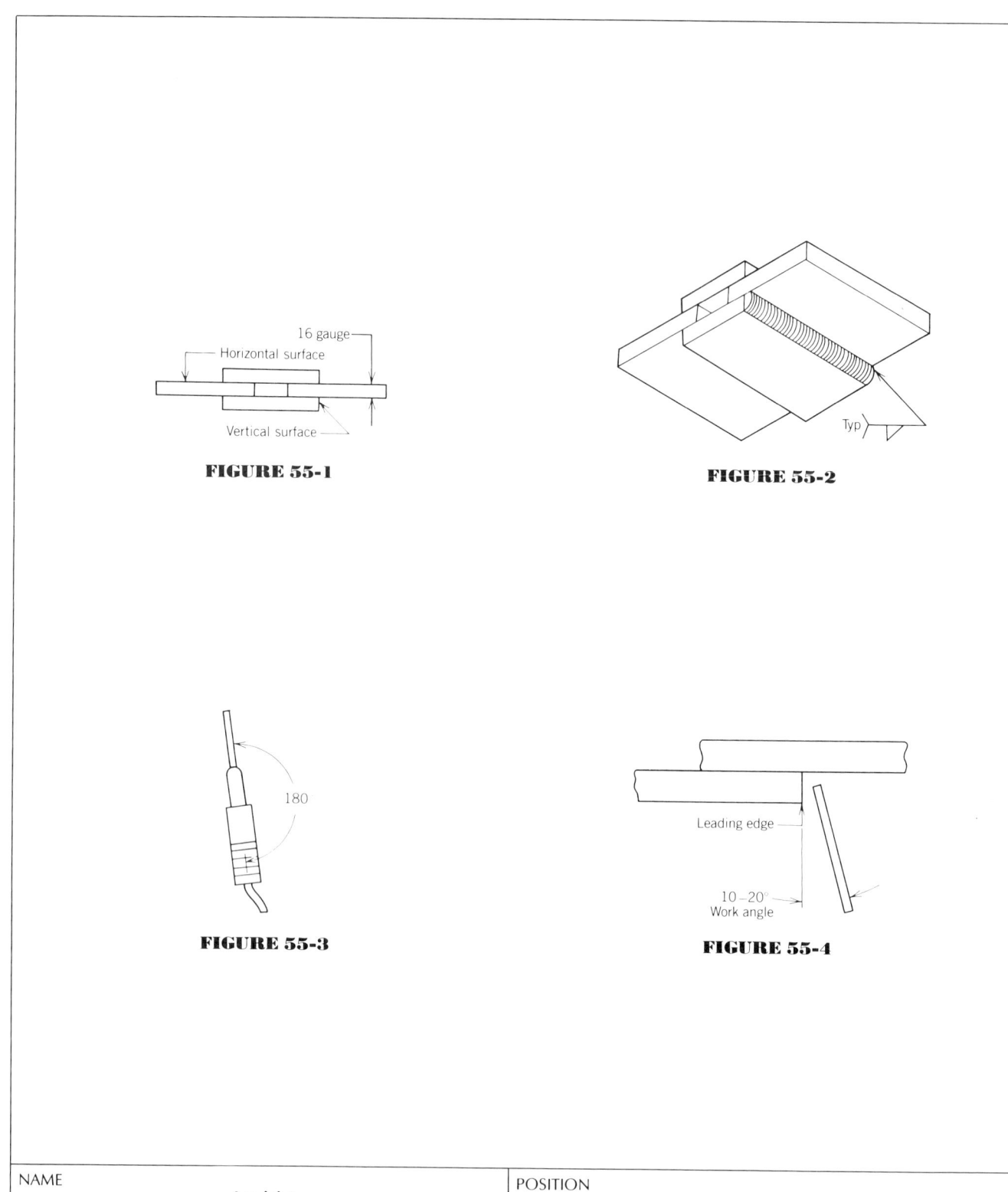

FIGURE 55-1

FIGURE 55-2

FIGURE 55-3

FIGURE 55-4

NAME	Lap joint	POSITION	Overhead
ELECTRODE E-6013	DIAMETER 3/32"	AMPERAGE	45-90 DCEN
MATERIAL	4 pcs. 16 gauge × 2" × 10" carbon steel		
PASSES Single	BEAD String	TIME (SEE INSTRUCTOR)	

Procedure 55

Lap Joint Fillet, 4F Position

OBJECTIVE

Upon completion of this lesson you should be able to weld overhead lap joints on sheet metal.

Text Reference: Section III, Lesson 6D; page 250.

PROCEDURE

1. Adjust the welding current.
2. Tack the weldment in the position shown in Figure 55-1. Clean the tacks and all intended weld areas.
3. Secure the plates in the 4F position as shown in Figure 55-2.
4. Place the electrode in the holder as shown in Figure 55-3.
5. Start at the left side of the joint with the electrode held at a 10- to 20-degree work angle off the vertical face of the joint as shown in Figure 55-4.
6. Favor the horizontal surface, making sure the leading edge and the vertical leg are fused. Do not burn the leading edge of the bottom sheet back more than approximately 1/16 in.
7. Use a slight "C" motion, or the whipping method, to control the puddle. Do not overweld by making too wide a bead. Do not pile excess weld metal on the surface. Guard against burn through in the last few inches.
8. Clean, cool, and examine the joint for appearance and soundness. The weld contour should be flat or slightly convex.
9. Weld the remaining three fillets as indicated by the symbol in Figure 55-2. "Typ" means all joints of the type indicated should be welded the same. Weld two joints using DCEN and two using alternating current.

124 SHIELDED METAL ARC WELDING PROCESS

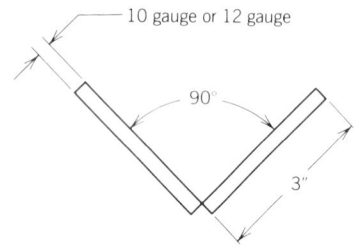

FIGURE 56-1

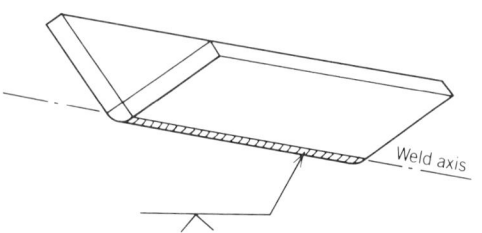

FIGURE 56-2

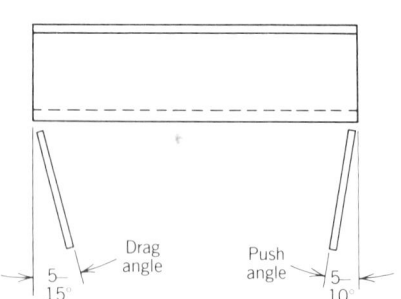

FIGURE 56-3

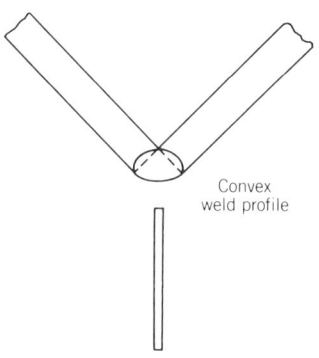

FIGURE 56-4

NAME	Corner joint	POSITION	Overhead
ELECTRODE E-6012	DIAMETER 1/8"	AMPERAGE	80–140 DCEN
MATERIAL	4 pcs. 10 or 12 gauge × 3" × 10" carbon steel		
PASSES Single	BEAD String	TIME (SEE INSTRUCTOR)	

Procedure 56

Outside-Corner Joint, 4G Position

OBJECTIVE

Upon completion of this lesson you should be able to weld outside-corner joints in the overhead position.

Text Reference: Section III, Lesson 6D; page 251.

PROCEDURE

1. Adjust the welding current—alternating current or direct current either polarity, however, direct current electrode negative is preferred, DCEN.
2. Tack the weldment in the position shown in Figure 56-1. Clean away the slag and all intended weld areas.
3. Secure the weldment in the 4G position as shown in Figure 56-2.
4. Place the electrode in the holder as in the previous lesson.
5. Start welding at the left side of the joint. Use a drag angle of 5 to 15 degrees as shown in Figure 56-3. Use a work angle perpendicular to the joint as shown in Figure 56-4.
6. Make sure the puddle covers the entire joint, overlapping the toes slightly. The metal edges will melt and become part of the joint.
7. Use the drag technique or a slight whipping motion to weld the joint. Make sure you obtain complete fusion. The weld metal should not penetrate through to the opposite side of the joint. Hold a medium arc length.
8. Maintain the electrode drag angle to about the last 1 ½ in., then you may have to switch to a push angle of 5 to 10 degrees. This is illustrated in Figure 56-3. This is necessary only if the joint is overheating and there is a possibility of melt through.
9. Clean, cool, and examine the weld for defects.
10. Check for proper bead contour (see Figure 56-4).

126 SHIELDED METAL ARC WELDING PROCESS

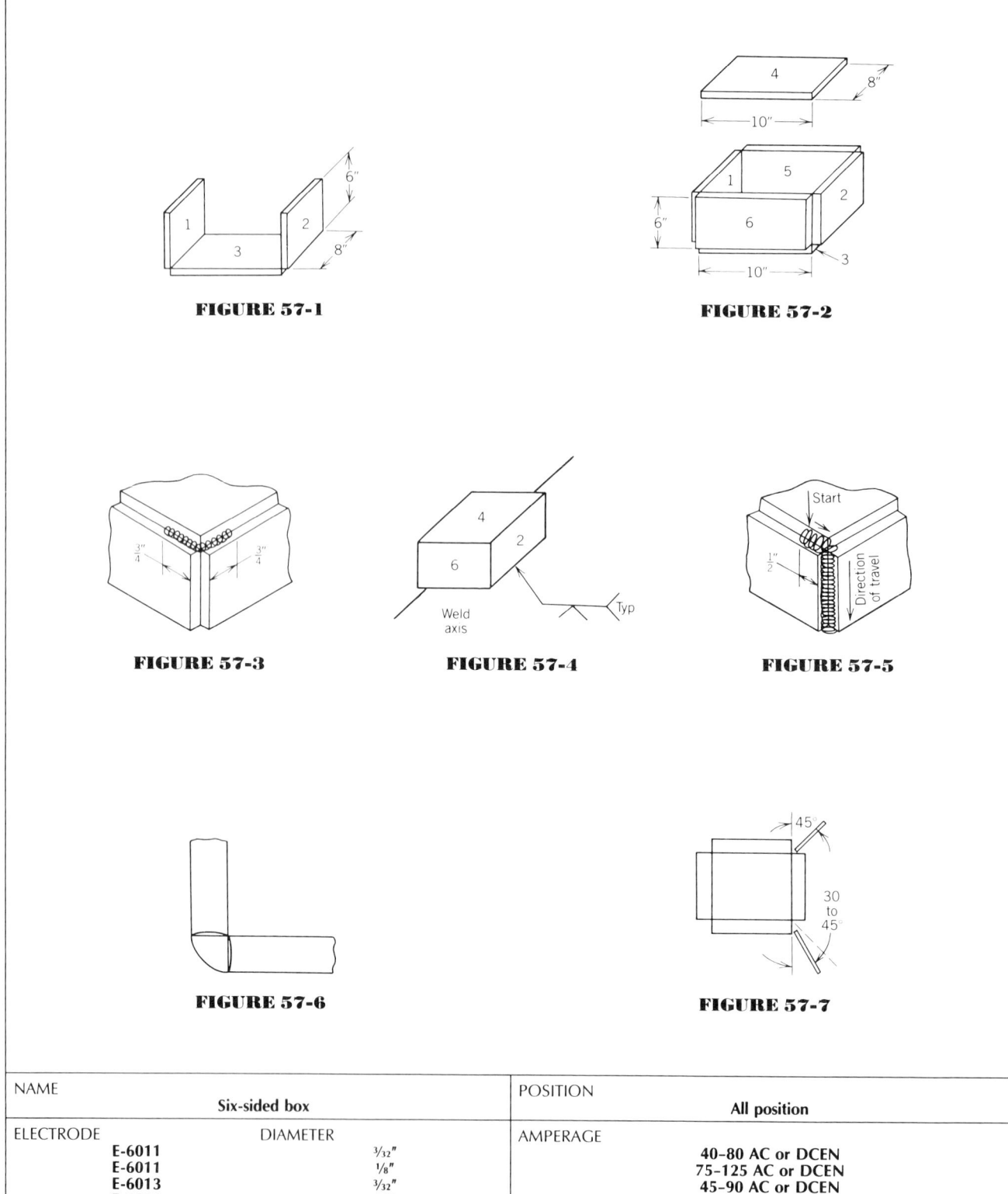

FIGURE 57-1

FIGURE 57-2

FIGURE 57-3

FIGURE 57-4

FIGURE 57-5

FIGURE 57-6

FIGURE 57-7

NAME			POSITION	
	Six-sided box			All position
ELECTRODE		DIAMETER	AMPERAGE	
	E-6011	$3/32''$		40–80 AC or DCEN
	E-6011	$1/8''$		75–125 AC or DCEN
	E-6013	$3/32''$		45–90 AC or DCEN
	E-6013	$1/8''$		80–130 AC or DCEN
MATERIAL				
	2 pcs. 10 gauge × 6″ × 8″ carbon steel			
	2 pcs. 10 gauge × 6″ × 10″ carbon steel			
	2 pcs. 10 gauge × 8″ × 10″ carbon steel			
PASSES		BEAD	TIME (SEE INSTRUCTOR)	
	Single		String	

Procedure 57

Welding a Six-Sided Box, in the 2G, 3G, and 4G Positions

OBJECTIVE

Upon completion of this lesson you should be able to weld a variety of joints using different electrodes to weld a closed-sheet metal box.

Text Reference: Section III, Lesson 6E; page 252.

PROCEDURE

1. Adjust the welding current as follows:

Electrode diameter Welding current (AC or direct current electrode negative)	Ampere Range
E-6011 3/32 in.	40 to 80
E-6011 1/8 in.	75 to 125
E-6013 3/32 in.	45 to 90
E-6013 1/8 in.	80 to 130

   Be sure all sheets are joined at a right angle. Use a try square to aid in positioning them. Clean all joints as in previous lessons. Maintain a close fit throughout.

2. Use a 3/32 in. electrode for the tacking procedure. Use the ballpeen hammer to close the joint openings.
3. Tack end pieces 1 and 2 to the bottom piece 3 as in Figure 57-1.
4. Tack side pieces 5 and 6 to form a hollow box as in Figure 57-2.
5. Tack top piece 4 to complete the box as in Figure 57-2. Check to be sure all the joints are fit properly. There should be no openings. Use the smaller diameter electrode to tack weld all corners for approximately 3/4 in. on each side of the vertical joint. This is shown in Figure 57-3.
6. Clean all the tacks and the intended weld areas. Place the weldment in position as in Figure 57-4. It must remain fixed in this position until all welding is completed.
7. Weld the four vertical joints. Use 1/8 in. diameter E-6011 electrodes on DCEN and the downhill technique. Start on the tack at the top, slightly to one side of the joint, as in Figure 57-5. Finish the same way at the bottom of the joint. This will allow you to overlap the connection with the remaining welds. It will also help eliminate poor connections at the junction of the welds. The contour of all welds is shown in Figure 57-6.
8. After welding all four vertical joints using this procedure clean and weld the bottom joint. Use the 1/8 in. diameter E-6013 electrode on DCEN. The work angle of the electrode should be approximately 30 to 40 degrees off the vertical leg as in Figure 57-7.
9. Complete the box by welding the top joint with the 1/8 in. diameter E-6011 electrode on DCEN. The work angle should be approximately 45 degrees off the horizontal leg of the joint. (See Figure 57-7.)
10. Clean all welds and allow to cool. Examine for any defects.

Shielded Metal Arc Welding of Pipe

The Procedures in this section will introduce you to the fundamentals of shielded metal arc welding of pipe. You will find the welding of pipe a new challenge that involves dexterity and a sharp perception of welding in three dimensions. A successful pipe welder exhibits exceptional ability to concentrate on the weld puddle as he or she guides the arc around the joint, all the while maintaining the proper electrode angle. You may find that some pipe welding situations will involve the use of mirrors due to inaccessibility to the pipe because of location of the pipe.

Complete the Procedures in this section in the sequence in which they appear. Only by mastering the basic techniques of pipe welding will you be able to successfully advance to more complex welding procedures.

GENERAL INSTRUCTIONS

The Procedures in this section will introduce you to techniques instrumental in becoming a successful pipe welder. You will need to review those skills involved in the procedures on oxyacetylene welding of pipe, presented earlier in the workbook.

MATERIALS AND EQUIPMENT

For this section the material and equipment will be as follows:

1. Welding goggles, number 5 or 6 lens
2. Safety glasses
3. Grinding goggles
4. Soapstone
5. Wraparound
6. Handgrinder or sander
7. Chipping hammer
8. Ballpeen hammer
9. Half-round bastard file, 10 in.
10. Wire brush

130 SHIELDED METAL ARC WELDING PROCESS

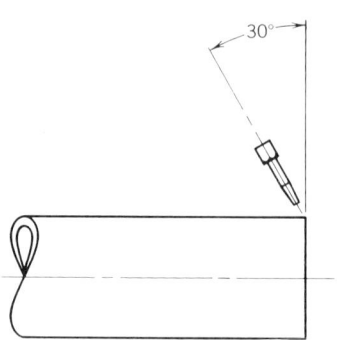

FIGURE 58-1

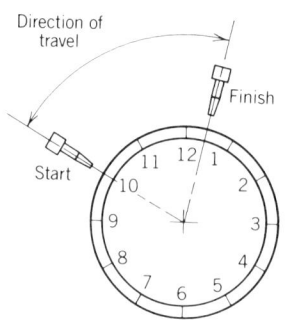

FIGURE 58-2

NAME Pipe preparation	POSITION Flat
ELECTRODE None　　DIAMETER	AMPERAGE
MATERIAL 1 pc. 4" or 6" schedule 40 pipe	
PASSES None　　BEAD None	TIME (SEE INSTRUCTOR)

Procedure 58

Beveling Pipe to Prepare It for Tests

OBJECTIVE

Upon completion of this lesson you should be able to prepare pipe for welding.

Text Reference: Section III, Lesson 7A; page 254.

PROCEDURE

1. Hammer the surface of the pipe slightly. Make sure you remove heavy rust deposits on the inside and outside surfaces.
2. Wire brush all loose residue and scale from the surfaces of the pipe.
3. Set up the oxy-fuel flame cutting unit you will use to prepare the pipe.
4. Using a wraparound, draw a line with the soapstone, about ¾ in. from the end of the pipe. (Make sure you use a well-sharpened soapstone.)
5. Place another line about 4 in. from the first.
6. If the pipe is long enough, and a pipe beveler is available, make cuts on both lines. If a pipe beveler is not available, use a hand torch.
7. If a hand torch is used, hold the torch at a 30-degree angle, with the torch tip within ⅛ in. or less from the surface of the pipe (see Figure 58-1).
8. Right-handed students should start cutting at approximately the 10 o'clock position. Then cut clockwise, stopping a little after 12 o'clock as shown in Figure 58-2. Do not attempt to make cuts longer than from 10 to 12:30. Left-handed students start at 2 and cut counterclockwise toward 12 o'clock.
9. Start the cut ¼ in. from the line, between the line and the scrap end of the pipe. After piercing the metal, make a cut to the line and begin to move along it slowly. Make sure you hold the torch tip at 30 degrees. The torch tip must always point at the center of the pipe.
10. When you reach the limit of travel (12 or 12:30), move the torch tip away from the line toward the end of the pipe, opening a small hole before you release the cutting oxygen lever. This will give you a new starting point when you resume the cut. It also eliminates gouging the bevel when you start the next cut.
11. Rotate the pipe counterclockwise and make the second cut. Continue to rotate the pipe after each cut until the full bevel is completed.
12. Tap the feather edge of the bevel lightly with the ballpeen hammer. Try to remove as much of the slag as possible. If the slag cannot be removed, you were moving too slow or your flame was set too high. Use a chipping hammer to remove the remainder of the slag.
13. Place the beveled pipe section in the vise. Then sand or file the remaining slag and scale from the beveled surface.
14. Use a sander or a file to remove the feather edge and form a root face.
15. Use the file to finish the root face. Do not forget to file all scale from outside of the pipe next to the beveled edge. Then use the rounded side of the file to clean inside the pipe.
16. After you have completed the bevel, cut off the pipe along the second line.

132 SHIELDED METAL ARC WELDING PROCESS

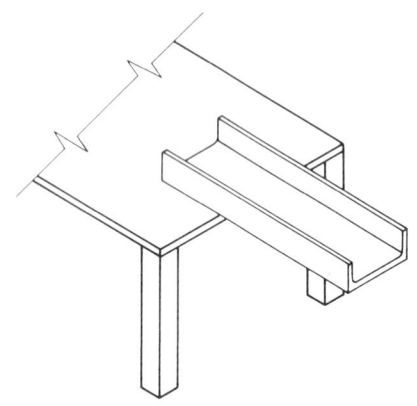

FIGURE 59-1

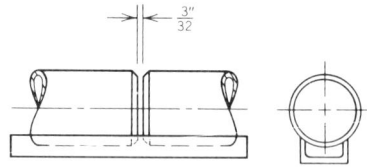

FIGURE 59-2

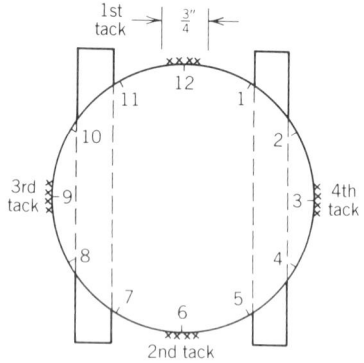

FIGURE 59-3

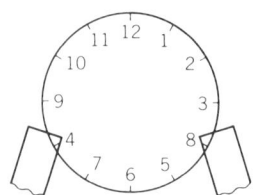

FIGURE 59-4

NAME			POSITION	
	Pipe fit and tack			Flat
ELECTRODE	DIAMETER		AMPERAGE	
E-6010		1/8"		75–125 DCEP
or				
E-6011		1/8"		75–125 AC or DCEP
MATERIAL				
2 pcs. 4" or 6" schedule 40 pipe from procedure 58				
2 spacers 3/32" × 5/8" × 6"				
1 pc. 3" or 4" channel 12" long				
PASSES	BEAD		TIME (SEE INSTRUCTOR)	
None		Tack		

Procedure 59

Alignment, Fit-Up, and Tacking Procedure

OBJECTIVE

Upon completion of this lesson you should be able to fit and tack pipe.

Text Reference: Section III, Lesson 7A; page 255.

PROCEDURE

1. Check to make sure that the outside surfaces of the pipes do not have any heavy coating or slag on them.
2. Tack or clamp the channel iron to a work table as shown in Figure 59-1.
3. Adjust the welding current.
4. Set the two pieces of pipe on the toes of the channel iron with the root faces almost touching as in Figure 59-2.
5. Place the two spacers between the root faces. Place one vertically, to run between 10 and 11 o'clock and 8 and 7 o'clock as shown in Figure 59-3.
6. The pipes must be pressed firmly together so the spacers cannot move. This will keep the root gap the same throughout the joint.
7. Place the first tack at 12 o'clock as shown in Figure 59-3. All tacks should be approximately ¾ in. long.
8. Remove the spacers and place them or wedges at 4 and 8 o'clock as shown in Figure 59-4. If you place them too deep into the root gap, it will be difficult to remove them. Rotate the pipe so you can tack the joint at 6 o'clock.
9. Place the second tack at 6 o'clock as shown in Figure 59-3, then remove the spacers.
10. Check to determine if the spacing at 3 and 9 o'clock are equal. If not, use the feather wedge just below 3 or 9 o'clock to even the spacing. Then place a tack just above the wedge. Remove the wedge. Then place it in the joint, on the opposite side, and tack weld just above the wedge.
11. Examine the joint. If the gap is not evenly spaced all around the pipe, or if the pipe is misaligned, cut the tacks out with a hacksaw and start from the beginning.

134 SHIELDED METAL ARC WELDING PROCESS

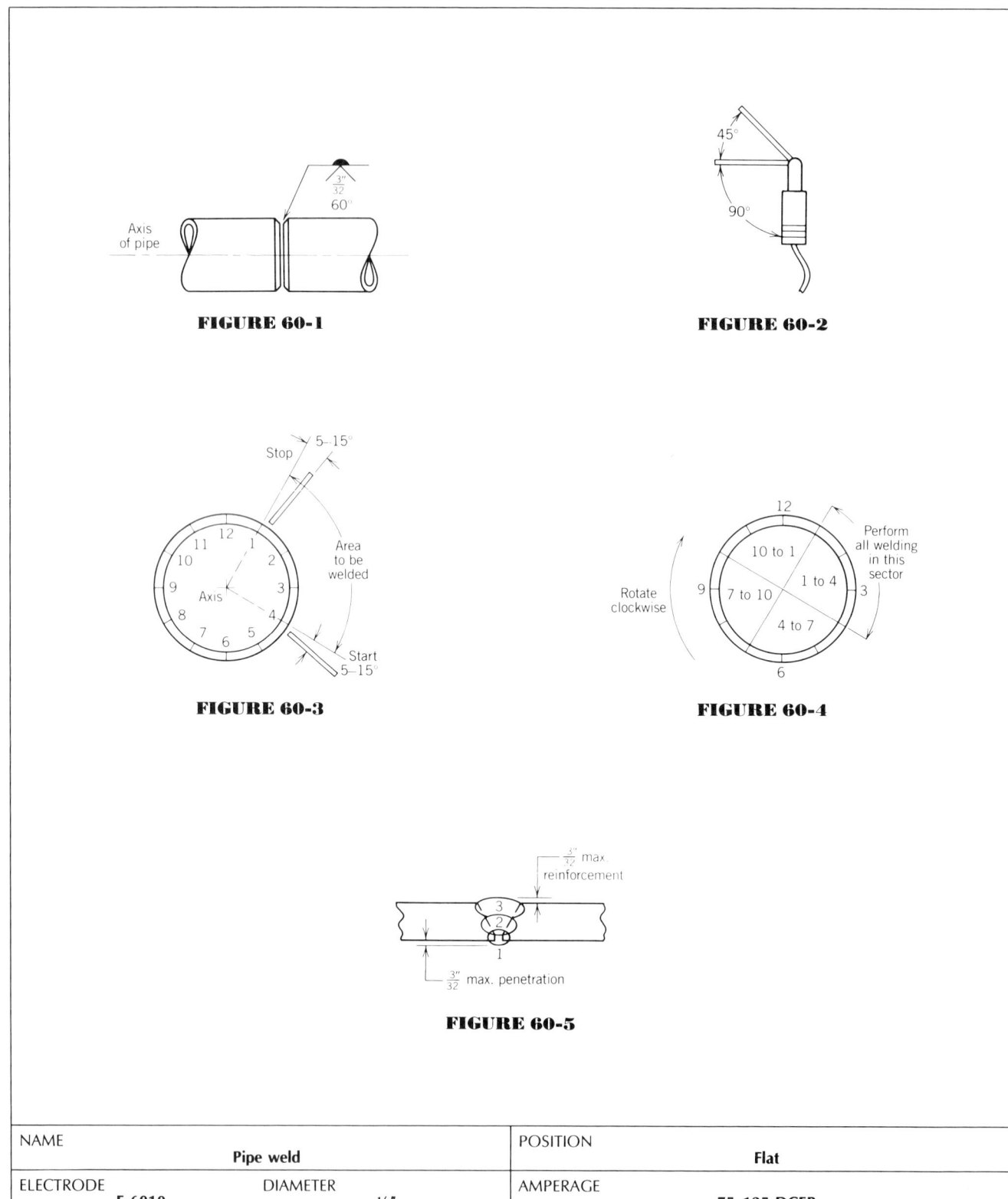

FIGURE 60-1

FIGURE 60-2

FIGURE 60-3

FIGURE 60-4

FIGURE 60-5

NAME	Pipe weld		POSITION	Flat
ELECTRODE	DIAMETER		AMPERAGE	
E-6010		1/8"		75-125 DCEP
or				
E-6011		1/8"		75-125 AC or DCEP
MATERIAL	2 pcs. 4" or 6" schedule 40 pipe			
PASSES Multiple	BEAD	Weave	TIME (SEE INSTRUCTOR)	

Procedure 60

Welding Pipe in the 1G Position, with Its Axis in the Horizontal Position

OBJECTIVE

Upon completion of this lesson you should be able to weld pipe one-quarter turn at a time.

Text Reference: Section III, Lesson 7B; page 257.

PROCEDURE

1. Prepare, fit, and tack the pipe.
2. Secure the pipe with its axis in the horizontal position as shown in Figure 60-1.
3. Adjust the welding current.
4. Place the electrode in the holder, as in Figure 60-2. Use a 90- or 45-degree angle away from the end of the holder.
5. Position yourself so you are at a 90-degree angle to the pipe. Be sure you are comfortable.
6. Strike the arc, on the bevel, at approximately 3 o'clock. Carry it down to 4 o'clock. Pause long enough for the root faces to melt away and a keyhole form. Then reverse your electrode direction.
7. Utilize the whipping method, as in welding plate in the vertical position, to run the first pass uphill. Use an electrode to push an angle 5 to 15 degrees upward as in Figure 60-3. Whip upward, taking care not to scar the surface of the pipe on either side of the V-groove. Stop when 1 o'clock is reached as shown in Figure 60-4. Clean thoroughly.
8. Turn the pipe clockwise toward you one-quarter of a turn. Then proceed in the same manner until the first pass is completed. Be sure to start the next electrode slightly below the crater.
9. The second pass (hot pass) and third pass (cover pass) can be welded with either the triangle motion or the alternate weave, as in vertical plate welding. Take care to pause at the sides of the joint. Burn out any entrapped slag and fill in any undesirable undercut.
10. Follow the same bead sequence. Adhere to the maximum root and face reinforcement as shown in Figure 60-5.
11. When you make the connection on completing the pass, be sure to overlap slightly. Break the arc by slowly drawing it away from the puddle.

136 SHIELDED METAL ARC WELDING PROCESS

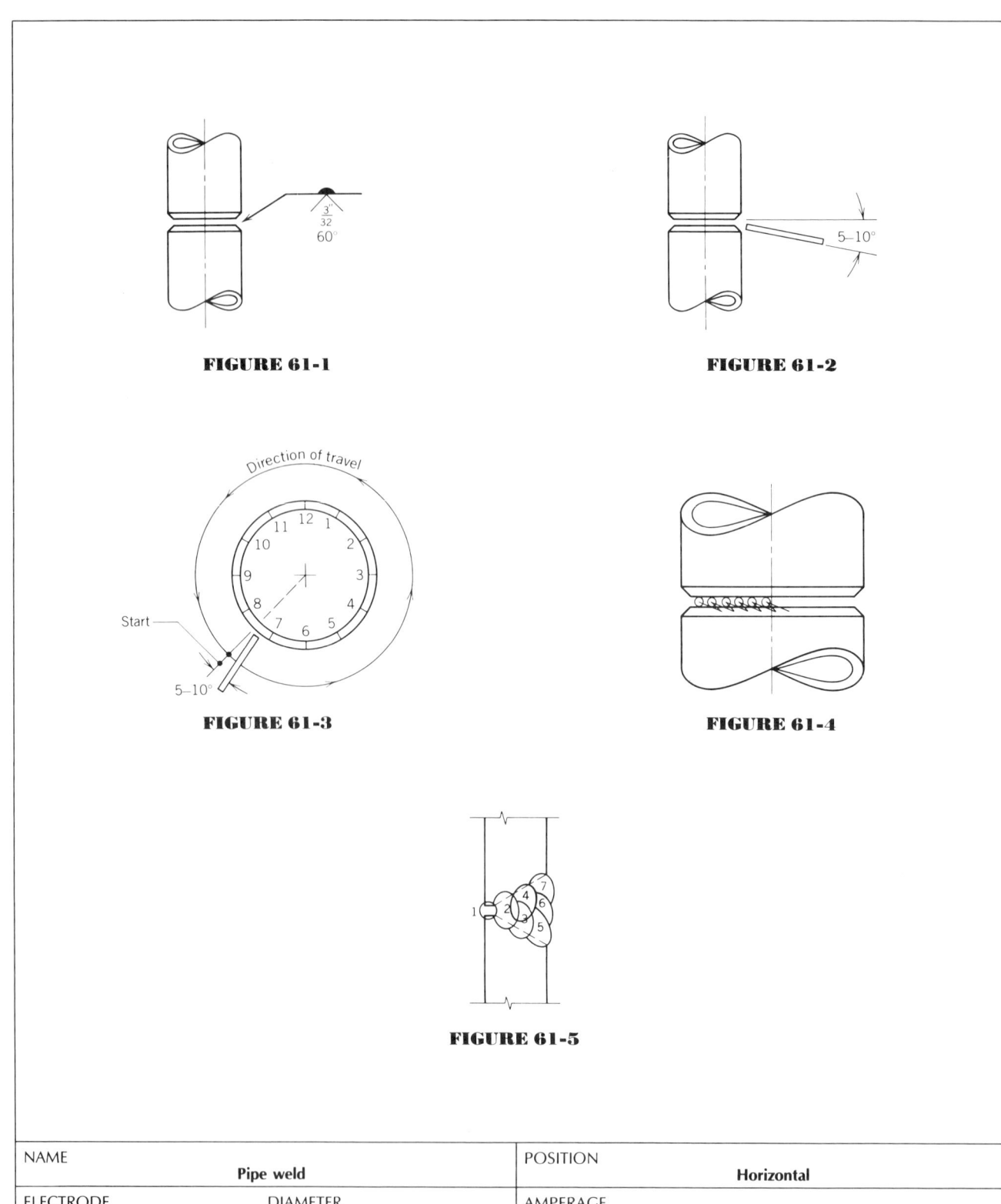

FIGURE 61-1

FIGURE 61-2

FIGURE 61-3

FIGURE 61-4

FIGURE 61-5

NAME	Pipe weld		POSITION	Horizontal
ELECTRODE	DIAMETER		AMPERAGE	
E-6010		1/8"		75–125 DCEP
or				
E-6011		1/8"		75–125 AC or DCEP
MATERIAL	2 pcs. 4" or 6" schedule 40 pipe			
PASSES	BEAD		TIME (SEE INSTRUCTOR)	
Multiple	Weave			

Procedure 61

Welding Pipe in the 2G Test Position, with Its Axis in the Vertical Position (E-6010 or E-6011 Electrodes)

OBJECTIVE

Upon completion of this lesson you should be able to weld pipe in the 2G position with E-6010 or E-6011 electrodes.

Text Reference: Section III, Lesson 7C; page 259.

PROCEDURE

1. Prepare the pipe and fit and tack.
2. Secure the pipe so its axis is vertical as in Figure 61-1.
3. Adjust the welding current.
4. Once the test has started the pipe may not be moved for any reason until the welding is completed. This is standard procedure when taking tests of this type.
5. Place the electrode in the holder, at a 90- or 45-degree angle, as in Lesson 7B.
6. The work angle of the electrode should be about 5 to 10 degrees below center of the joint as in Figure 61-2.
7. Use a slight drag angle of 5 to 10 degrees from the axis of the pipe as shown in Figure 61-3.
8. Strike an arc and get a keyhole started. If the upper root face burns away excessively, decrease the electrode work angle. You can do this by raising your hand and reducing the angle.
9. As shown in Figure 61-4 use the whipping motion to carry the arc onto the lower beveled surface. When you return to the keyhole, wash the metal into the root, then pause just long enough to fill the keyhole and reopen it again.
10. Make sure that you hold a long arc when starting a new electrode. This heats the metal where you left off welding so that you get the desired penetration when you enter the keyhole. You must obtain 100 percent penetration at these connections. Otherwise, you will leave voids or unwelded spaces. These will cause a weld to fail in service or during the testing procedure.
11. Clean the root pass thoroughly. Then run a stringer wide enough to cover the root pass. Make sure the weld toes are fused properly to prevent slag inclusions or undercut.
12. The next pass consists of stringers 3 and 4 as shown in Figure 61-5. These two stringers will be similar to those used in welding plate in the horizontal position. Weld the lower stringer first.

Remember: Only two-thirds of the lower stringer should be visible when the pass is completed.

13. The last pass consists of stringers 5, 6, and 7 as shown in Figure 61-5. If you used heavy beads on the previous passes, this pass may not be necessary.
14. Apply the three stringers as in horizontal plate welding. Be careful to avoid undercut on the top edge. You can use a slight push angle on the last stringer if undercut is a problem.

138 SHIELDED METAL ARC WELDING PROCESS

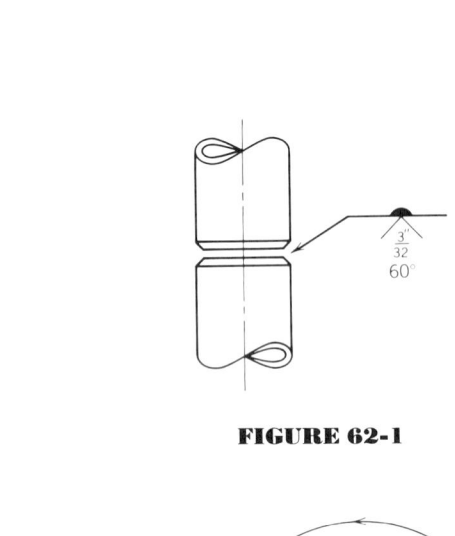

FIGURE 62-1 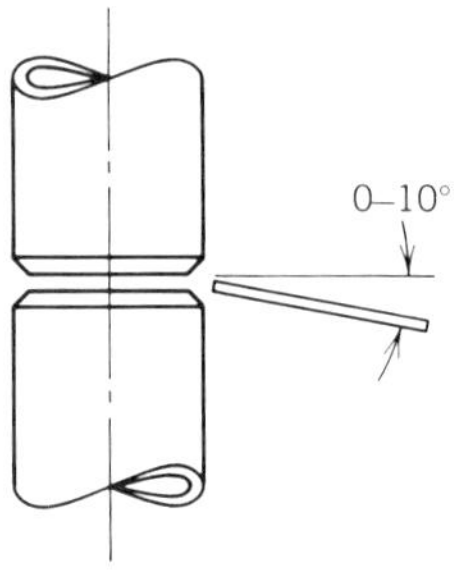   **FIGURE 62-2**

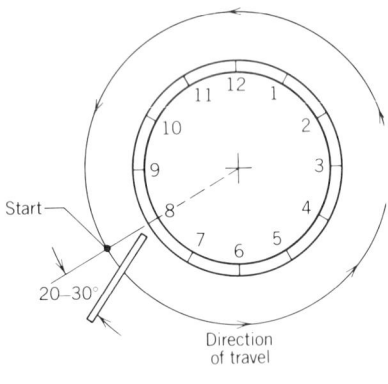

FIGURE 62-3

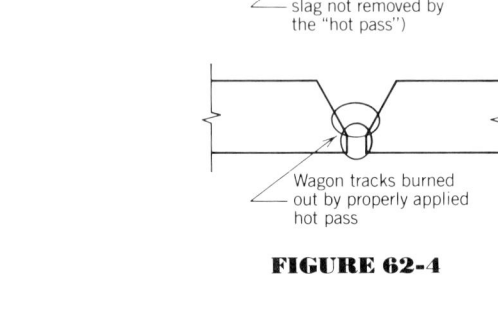

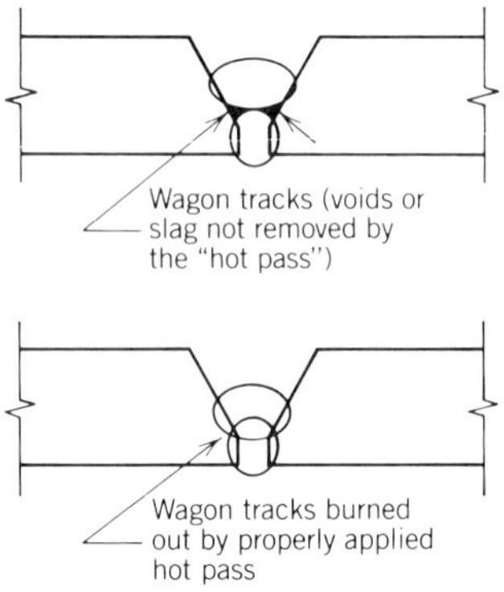

FIGURE 62-4

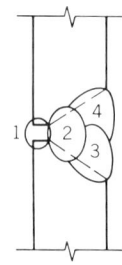

FIGURE 62-5

NAME	Pipe weld		POSITION	Horizontal
ELECTRODE	E-6010 or	DIAMETER 1/8"	AMPERAGE	75-125 DCEP
	E-6011	1/8"		75-125 AC or DCEP
	E-7018	1/8"		115-165 DCEP
MATERIAL	2 pcs. 4" or 6" schedule 40 pipe			
PASSES Multiple	BEAD Weave		TIME (SEE INSTRUCTOR)	

Procedure 62

Welding Pipe in the 2G Test Position, with Its Axis in the Vertical Position (E-6010 or E-6011 and E-7018 Electrodes)

OBJECTIVE

Upon completion of this lesson you should be able to weld pipe in the 2G position using E-6010 or E-6011 electrodes for the root pass, and E-7018 electrodes for the fill pass.

Text Reference: Section III, Lesson 7C; page 260.

PROCEDURE

1. Prepare the pipe and fit and tack.
2. Adjust the welding current.
3. Secure the weldment in the 2G position as shown in Figure 62-1.
4. Place the E-6010 or E-6011 electrode in the electrode holder at a 45-degree angle.
5. Assume a comfortable position. Lean against the pipe or balance yourself.
6. Strike an arc on the beveled surface and move quickly down into the root gap. Hold the electrode with a 20- to 30-degree drag angle and a 0- to 10-degree work angle, upward into the root, as shown in Figures 62-2 and 62-3.
7. Obtain penetration and complete the root pass. Hold a long arc on restarts. Be sure to "burn in" when making the connection.
8. Clean the weld area thoroughly. Change over to the E-7018 electrode. Increase the amperage slightly as recommended and try the arc on scrap metal.
9. Strike an arc. Change to a drag angle of 5 to 15 degrees, but use the same work angle as in Figure 62-2. Hold a short arc and do not whip out of the puddle. Just move the electrode along at a steady rate and pull the puddle with it.
10. The electrode may be oscillated, or moved slightly within the confines of the puddle. Keep the oscillations to a minimum, however.
11. This second pass is a stringer bead. Pay close attention; keep the crater deep enough and hot enough to fuse the sides of the joint and the root pass thoroughly. Welders sometimes call this "burning out the wagon tracks" with the "hot pass." This is illustrated in Figure 62-4.
12. Clean the weld thoroughly. Paying close attention to removing any slag that may remain at the toes of the weld. Cover this pass with at least two more stringers as in Figure 62-5.
13. Clean each pass carefully. Be sure to hold normal arc lengths on restarts. The arc should be struck ½ in. ahead of the crater and slowly moved back to the crater. This will reweld the arc strike areas as the weld progresses.
14. The last stringer is critical. Watch the upper left-hand edge of the puddle to be sure that you are filling in all undercut.

140 SHIELDED METAL ARC WELDING PROCESS

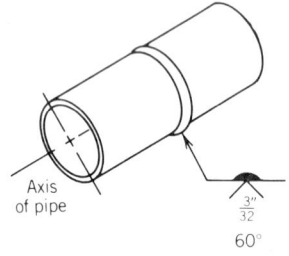

FIGURE 63-1

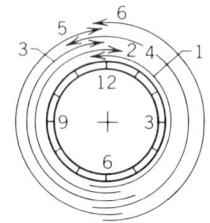

FIGURE 63-2

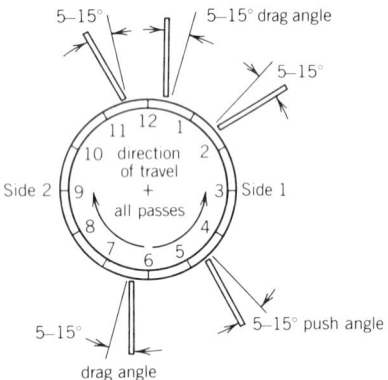

FIGURE 63-3

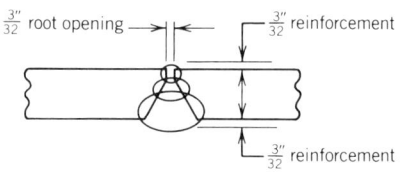

FIGURE 63-4

NAME	Pipe weld	POSITION	Horizontal fixed
ELECTRODE E-6010 or E-6011	DIAMETER 1/8" 1/8"	AMPERAGE	75-125 DCEP 75-125 AC or DCEP
MATERIAL	2 pcs. 4" or 6" schedule 40 pipe 2 spacers 3/32" × 5/8" × 6"		
PASSES Multiple	BEAD Weave	TIME (SEE INSTRUCTOR)	

SHIELDED METAL ARC WELDING OF PIPE

Procedure 63

Welding Pipe in the 5G Position Uphill Passes (E-6010 or E-6011 Electrodes)

OBJECTIVE

Upon completion of this lesson you should be able to weld pipe in the 5-G position.

Text Reference: Section III, Lesson 7D; page 261.

PROCEDURE

1. Prepare, fit, and tack the pipe.
2. Secure the pipe in the 5G (horizontal fixed) position as in Figure 63-1. Choose a height that will allow you to reach the top and bottom of the joint comfortably.
3. Adjust the welding current.
4. Place the electrode in the holder at a 90- or 45-degree angle, whichever is more comfortable for you.
5. Get into as comfortable position as possible. Lean your shoulder or upper arm against the pipe.
6. Strike an arc on the bottom beveled surface at about the 6:30 position. Weld counterclockwise through the 6 o'clock position, up the side one of the pipe. This is shown Figure 63-2.
7. Use an electrode work angle of 90 degrees and a drag angle of about 5 to 15 degrees as shown in Figure 63-3.
8. Maintain a constant drag angle. Hold a close arc, similar to the drag technique, but without the electrode touching the pipe. Keep the arc visible while you whip the bead uphill toward 5 o'clock. Keep the keyhole open.
9. At slightly past 5 o'clock switch from a drag to a 5- to 15-degree push angle. Continue to weld upward as shown in Figure 63-3.
10. Use the same whipping technique as in vertical welding of the grooves in plate.

Remember: Maintain the slight push angle until you reach between the 1 and 2 o'clock position. At this point switch to a 5- to 15-degree drag angle, continue welding until 12:30 as in Figure 63-3.

11. Clean and inspect the weld. Move to side two of the pipe. (See the weld progression shown in Figure 63-2.) Now make the second bead.

Remember: Hold a long arc when overlapping starts and stops on the first bead. Clean and inspect the second bead.

12. Return to side two. Start the second pass (third bead) by welding uphill as shown in Figure 63-2. Use the weave method. You can start by moving the electrode from side to side; then continue unless the puddle becomes too fluid. If that is the case, switch to the triangle motion. The triangle motion may become necessary in the area of 8 to 10 o'clock. Maintain the electrode drag and push angles as shown in Figure 63-3. Clean and inspect the bead.
13. Move to side one. Start bead four and complete the second pass. Hold a long arc on the starts and stops. Clean and inspect the completed pass. See the bead sequence shown in Figure 63-4.
14. Start the final (cap) pass on side two. Begin at approximately 6 o'clock and weld uphill to 12 o'clock using the weave technique. Keep the weld width complete and the reinforcement to a minimum. (See Figure 63-4.)
15. Weld side one of the cap pass. Make sure to overlap at 6 and 12 o'clock.
16. Clean and examine your weld for undercut, gas pockets, slag holes on the surface, penetration, width, and reinforcement.

142 SHIELDED METAL ARC WELDING PROCESS

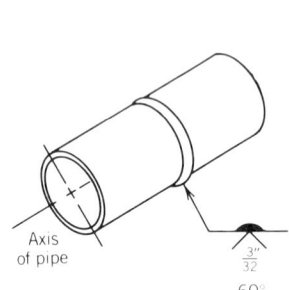

FIGURE 64-1

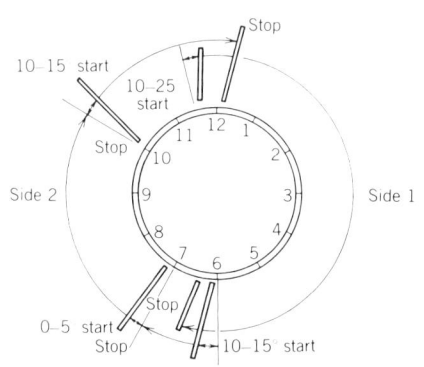

FIGURE 64-2

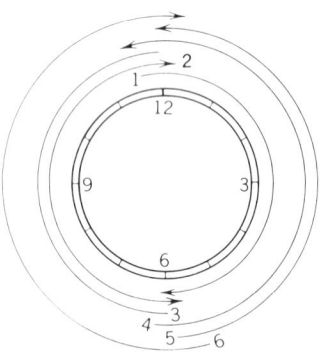

FIGURE 64-3

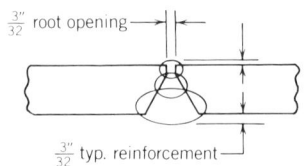

FIGURE 64-4

NAME	Pipe weld		POSITION	Horizontal fixed
ELECTRODE	DIAMETER		AMPERAGE	
E-6010		1/8"		75-125 DCEP
or				
E-6011		1/8"		75-125 AC or DCEP
E-7018		1/8"		115-165 DCEP
MATERIAL	2 pcs. 4" or 6" schedule 40 pipe 2 spacers 1/16" × 5/8" × 6"			
PASSES Multiple	BEAD Weave		TIME (SEE INSTRUCTOR)	

Procedure 64

Welding Pipe in the 5G Position First Pass Downhill (E-6010 or E-6011 Electrodes) Second and Out Uphill (E-7018 Electrodes)

OBJECTIVE

Upon completion of this lesson you should be able to weld pipe in the 5G position.

Text Reference: Section III, Lesson 7D; page 262.

PROCEDURE

1. Prepare and fit the pipe as in previous lessons.
2. Secure the pipe in the 5G (horizontal fixed) position as in Figure 64-1 at a comfortable welding height, to reach the top and bottom of the joint.
3. Adjust the welding current.
4. Place the electrode in the holder at a 90- or 45-degree angle, whichever is more comfortable for you.
5. As in lesson 63, get into as comfortable position as possible. Strike an arc on the beveled surface. Begin at about 11:30 position as shown in Figure 64-2. Weld clockwise, using the downhill technique as shown in Figure 64-3.
6. Keep the electrode work angle at 90 degrees to the surface of the pipe. The drag angle should be from 10 to 25 degrees, depending on the amount of penetration you obtain. If the drag angle is too large, stop and increase your amperage. This will give you more penetration and reduce the required electrode drag angle. As shown in Figure 64-3, weld downhill.
7. Stop just past 6 o'clock, as shown in Figure 64-2. Clean the bead and move to side two. Complete the second downhill root pass. Be sure to properly overlap the first bead.
8. Second pass and out use E-7018 low hydrogen electrodes. Use a drag angle of 10 to 15 degrees. Start the third bead somewhere near 5 to 7 o'clock, taking care to overlap and tie in the connection between the two root beads. Weld upward on side two. Use a slight weave. Take care not to leave the puddle or oscillate excessively.
9. As the weld passes 7 o'clock, gradually change from a drag angle of 0 to 5 degrees. Continue until 10 o'clock is reached. Then switch to a drag angle of 10 to 15 degrees. Stop at about 12 o'clock.
10. Clean and move to side one. Weld the fourth bead uphill. Hold a long arc on the starts and stops to make good tie in connections.
11. After cleaning use uphill beads to weld the finish pass. Start on side one. Do not weave the pass any wider than 3/32 in. past the edge of the bevels. As shown in Figure 64-4 keep the reinforcement to approximately 3/32 in. Beads with low hydrogen electrodes should be no wider than three electrode diameters. If the groove is wider than this, it would be better to finish the joint with stringers.
12. Cool and clean the completed weld. Examine for undercut, surface defects, bead shape, width, and appearance of the weld and reinforcement.

144 SHIELDED METAL ARC WELDING PROCESS

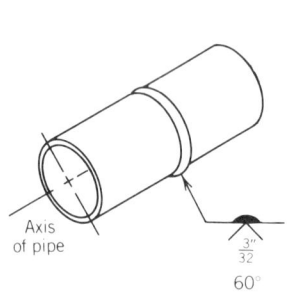

FIGURE 65-1

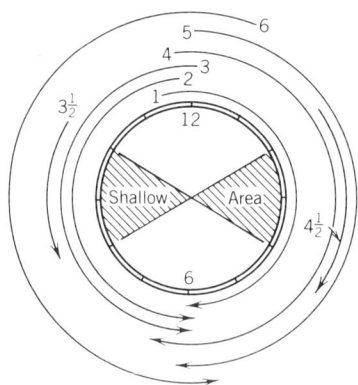

FIGURE 65-2

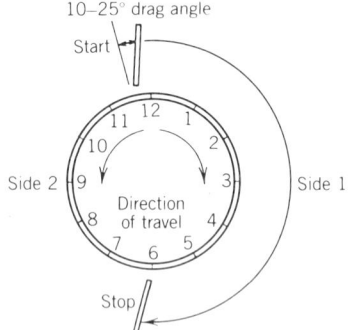

FIGURE 65-3

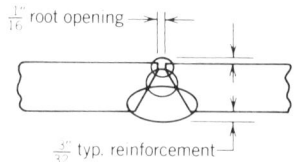

FIGURE 65-4

NAME	Pipe weld	POSITION	Horizontal fixed
ELECTRODE E-6010 E-6011 E-6010 E-6011	DIAMETER 1/8" 1/8" 5/32" 5/32"	AMPERAGE	75-125 DCEP 75-125 AC or DCEP 110-170 DCEP 110-170 AC or DCEP
MATERIAL	2 pcs. 4" or 6" schedule 40 pipe 2 spacers 1/16" × 5/8" × 6"		
PASSES Multiple	BEAD Weave	TIME (SEE INSTRUCTOR)	

Procedure 65

Welding Pipe in the 5G Position Downhill Passes (E-6010 or E-6011 Electrodes)

OBJECTIVE

Upon completion of this lesson you should be able to weld pipe in the 5G position.

Text Reference: Section III, Lessson 7D; page 263.

PROCEDURE

1. Prepare and fit the pipe as in previous lessons.
2. Secure the pipe in the 5G (horizontal fixed) position, as shown in Figure 65-1. Use a comfortable height so you can reach both top and bottom of the joint.
3. Adjust the welding current.
4. Place the electrode in the holder at a 90- or 45-degree angle, whichever is more comfortable for you.
5. Assume a comfortable welding position. Drag the first pass downhill, on side one. Weld from 11:30 to approximately 6 o'clock and as shown in Figures 65-2 and 65-3. Maintain a 90-degree work angle and 10- to 25-degree drag angle.
6. Clean the bead and weld side two of the root pass as shown in Figure 65-2. Be sure you are obtaining the correct melt through or penetration.
7. Start the second pass at the top of side two. The electrode angles remain the same as on the root pass, except the drag angle can be reduced to zero between 7 o'clock and the bottom on side two, and 5 o'clock and the bottom on side one.
8. Clean weld bead three and complete the second pass by welding bead four on side one as in Figure 65-2. Be sure to overlap at the start and stop.
9. Clean the second pass thoroughly, then examine it for evenness of deposit. You should have deposited the same amount of weld metal throughout the entire joint. Then finish off with a cap pass using the E-6010 or E-6011 $\frac{5}{32}$ in. electrode. (See Figure 65-4.)
10. Frequently the weld is shallow between 2 to 4 and 8 to 10 o'clock. Then, as shown in Figure 65-2, you must run short beads between these points. These extra beads are shown as 3½ and 4½ in Figure 65-2. These beads will allow you to weld a uniform finish pass around the entire joint.

146 SHIELDED METAL ARC WELDING PROCESS

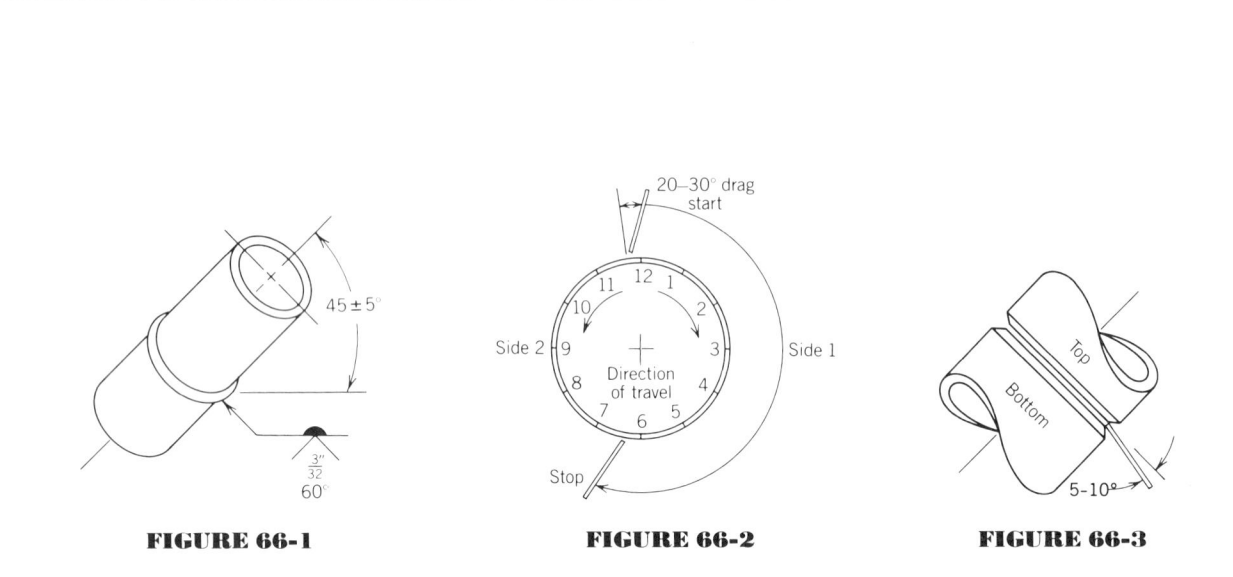

FIGURE 66-1 **FIGURE 66-2** **FIGURE 66-3**

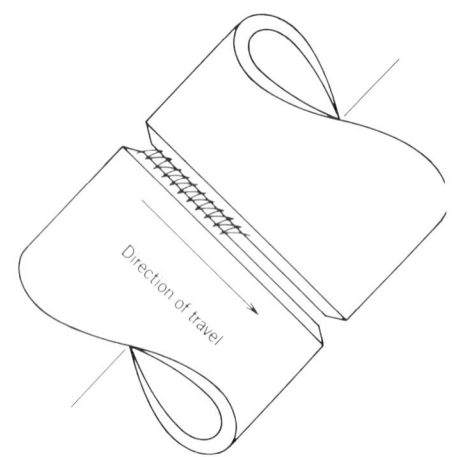

FIGURE 66-4

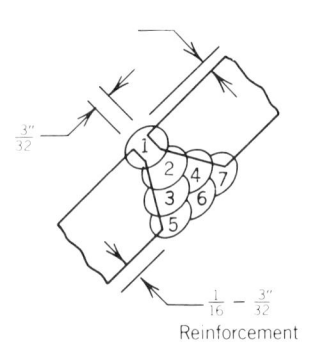

FIGURE 66-5

NAME	Pipe weld	POSITION	45 degree incline 6G
ELECTRODE E-6010 or E-6011	DIAMETER 1/8" 1/8"	AMPERAGE	75-125 DCEP 75-125 AC or DCEP
MATERIAL	2 pcs. 4" or 6" schedule 40 pipe 2 spacers 3/32" × 5/8" × 6"		
PASSES Multiple	BEAD Weave	TIME (SEE INSTRUCTOR)	

Procedure 66

Welding Pipe in the 6G Position Downhill Passes (E-6010 or E-6011 Electrodes)

OBJECTIVE

Upon completion of this lesson you should be able to weld pipe in the 6G position.

Text Reference: Section III, Lesson 7E; page 265.

PROCEDURE

1. Prepare and fit the pipe as in previous lessons.
2. Secure the pipe in the 6G position as shown in Figure 66-1. Use a height that is comfortable for you to reach.
3. Adjust the welding current.
4. Place the electrode in the holder at a 90-degree angle.
5. Position yourself as in previous lessons.
6. Strike the arc on the beveled surface of the joint, between 11:30 and 12 o'clock. Use a drag angle of 20 to 30 degrees to the surface of the pipe as shown in Figure 66-2. Maintain this angle for the entire first pass.
7. Maintain a work angle between 0 to 10 degrees off the perpendicular as shown in Figure 66-3. The angle you use will be determined by the way the root faces melt. Both root faces may melt evenly, or one may tend to melt faster and burn away. If one face melts faster, angle the electrode toward it. This forces weld metal into the area so the joint may fuse uniformly.
8. As you weld keep a constant pressure on the electrode. Keep lowering your hand and the electrode holder. This maintains the proper drag angle as the electrode burns away while you move along the changing contour of the pipe. Do not burn the electrode any shorter than 2½ to 3 in. It may overheat and cause the electrode to stick to the joint.
9. Before starting a second electrode the completed portion of the weld should be thoroughly cleaned. One electrode should be enough for you to complete the first pass, from 11:30 to 6:30.
10. Start the second half of the root pass on the opposite side of the pipe, side 2. Overlap the first bead by approximately ¾ in. Hold a long arc until the root opening is reached. Then force the electrode gently into the joint until the arc almost disappears. When the electrode is inside the joint, a harsh sound is heard. This sound is the best indication that you are obtaining penetration.
11. When you reach the end of the root opening at the bottom of the pipe, pull out to a normal arc length. Continue to weld until you overlap the first bead at least ¾ in.
12. Clean the pass thoroughly before you start the second, or "hot pass." (Use your chipping hammer, the saw blade, and wire brush.) Begin on side one. Start welding somewhere between 12 and 1 o'clock. Be sure to stagger the starts and stops.
13. For the hot pass, and all other passes, the electrode drag angle should be between 10 to 20 degrees. Also, you should always aim the electrode at a point slightly off the center of the pipe as in Figure 66-2. This amount of electrode angle tends to push the molten pool in the direction you wish. It also gives you better control of the weld deposit.
14. The hot pass not only deposits filler metal, it "burns out" any trapped slag in the toes of the root pass. Make sure you set the current high enough so the arc melts the crown of the root pass. When the puddle is concave and fluid, you probably are obtaining a good bead. Use a slight weaving motion as shown in Figure 66-4.
15. Do not pause as long at the joint sides as in uphill welding. Pausing will cause the center of the weld to sag; however, you should pause long enough to fuse the toes of the joint properly.
16. Repeat this on side two. Remember to overlap at least ¾ in. at the start and end of the preceding pass.
17. The next pass will be comprised of two stringer beads as shown in Figure 66-5. Run the first one on the lower part of the joint. Use the same electrode manipulation as you used for the hot pass.
18. Do not start both stringers at the same point. Stringers must start and stop on a staggered basis. Watch out for undercut on the upper stringer.
19. Clean the pass completely. Finish the joint with a three stringer cap pass as shown in 66-5.

The number of passes required to complete the pipe joint depends on the wall thickness and size of the stringers. Sometimes you can complete the joint by depositing a root pass, a hot pass, and a cap pass of two stringers.

148 SHIELDED METAL ARC WELDING PROCESS

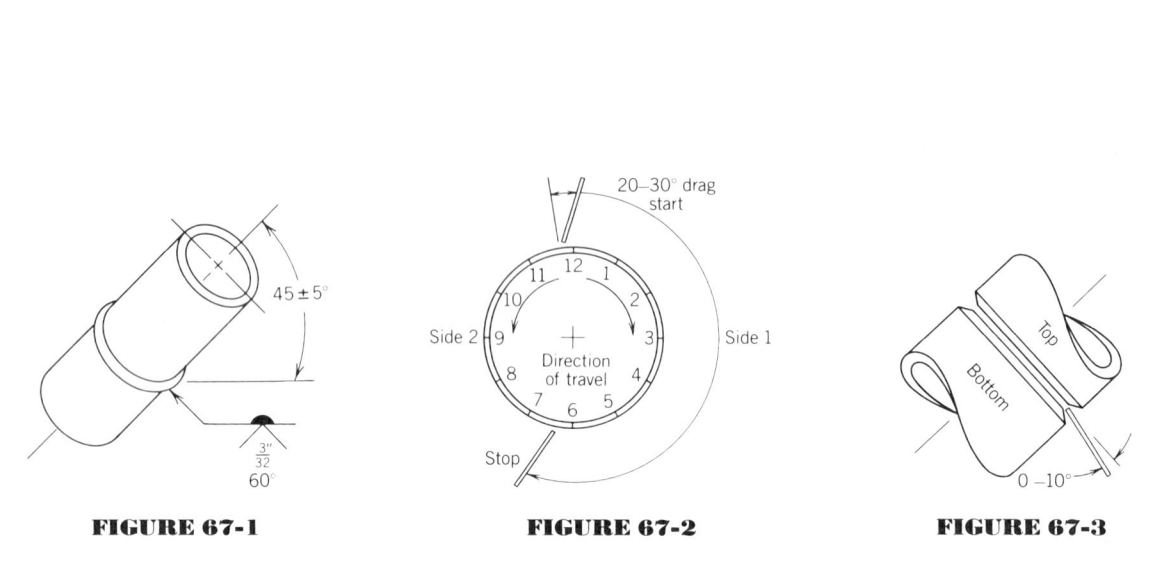

FIGURE 67-1 **FIGURE 67-2** **FIGURE 67-3**

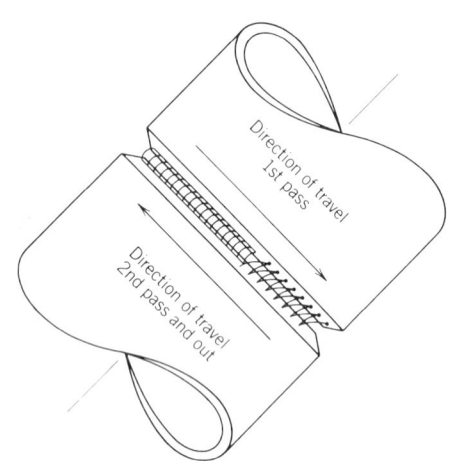

FIGURE 67-4

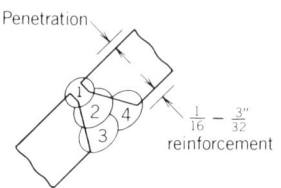

FIGURE 67-5

NAME	Pipe weld	POSITION	45 degree incline 6G
ELECTRODE	DIAMETER	AMPERAGE	
E-6010 or	1/8″		75-125 DCEP
E-6011	1/8″		75-125 AC or DCEP
E-7018	1/8″		115-116 DCEP
MATERIAL	2 pcs. 4″ or 6″ schedule 40 pipe 2 spacers 1/16″ × 5/8″ × 6″		
PASSES Multiple	BEAD Weave	TIME (SEE INSTRUCTOR)	

Procedure 67

Welding Pipe in the 6G Position First Pass Downhill (E-6010 and E-6011 Electrodes), Second and Out Uphill (E-7018 Electrode)

OBJECTIVE

Upon completion of this lesson you should be able to weld pipe in the 6G position.

Text Reference: Section III, Lesson 7E; page 266.

PROCEDURE

1. Prepare and fit the pipe as in previous lessons.
2. Secure the pipe in position as shown in Figure 67-1.
3. Adjust the welding current.
4. Drag the first pass with E-6010 or E-6011 electrodes. Make sure to use the electrode angles indicated in Figures 67-2 and 67-3.
5. Clean the first pass thoroughly, then switch to E-7018 electrodes. Adjust your amperage on scrap plate.
6. The iron powder in the coating of the E-7018 low hydrogen electrode puts more filler metal in the deposit than E-6010 or E-6011 electrodes. This means it will take fewer passes to fill the joint.
7. Start the second pass near the bottom of the pipe, between 5 and 6 o'clock. Run a stringer bead toward the top of side 2. Use a slight weaving motion as shown in Figure 67-4. Do not oscillate the electrode excessively. Do not whip or remove the electrode from the puddle at any time. Use an electrode drag angle of 0 to 5 degrees.
8. Clean the bead thoroughly, using the slag hammer and wire brush. Remember that the slag formed by low hydrogen electrodes is difficult to remove. It is almost impossible to burn out with the next pass.
9. Weld the other side of the pipe. Be careful of the overlap at both the start and stop. Follow the bead sequence shown in Figure 67-5.
10. The finish pass is run with stringer beads three and four. The lower stringer should be welded first. Use the same electrode angles and electrode motion as before. Be sure to stagger the starts and stops.

GAS TUNGSTEN ARC WELDING PROCESS

Gas Tungsten Arc Welding of Carbon Steel

In the following procedures you will develop your skills in gas tungsten arc welding. You will learn the fundamental techniques by welding on carbon steel. Carbon steel is readily available and is considerably less expensive than stainless steel or aluminum.

In later sections, you will weld aluminum, using much of what you learned on carbon steel. If you master gas tungsten arc welding of carbon steel and aluminum, you will be able to weld other materials with just a little additional instruction and practice.

GENERAL INSTRUCTIONS

The welding procedures will specify amperage ranges. If your equipment has remote amperage control foot pedal capabilities you may wish to use it, however, it is not required. You will find many cases in industry where foot pedals cannot be used because of the position or size of the weldment. It would be good to practice welding without remote current control so you will be better prepared for these situations. High frequency will be specified for ease in starting the arc. The arc can be initiated by using a touch start technique, but this is not a preferred method since it often results in causing a small portion of the end of the tungsten to break off and remain in the weld. The type of torch you use will depend on the equipment available in your shop. Air-cooled torches are limited in capacity. Water-cooled torches are preferred.

1. Use both hands when welding with filler wire addition, and one foot when using remote amperage control pedals.
2. Find a comfortable working stance. Remember it is difficult to produce sound quality welds if you are in an awkward or uncomfortable position.
3. Hold a short and steady arc. This means holding the torch as steady as possible. Few of us are steady enough to do that freehand. You should look for some way to support the hand you use to hold the torch while welding. Hold the torch in a way that is most comfortable to you. Some hold the torch as they would a pencil, while others hold it like an electrode holder.
4. Hold the filler wire between your thumb and forefinger with 10 to 12 in. sticking out.

MATERIALS AND EQUIPMENT

For this section the material and equipment will be as follows:

1. Welding shield with appropriate filter lens
2. Safety glasses
3. Short cuff leather gloves
4. Appropriate welding clothing
5. Wire brush
6. Pliers

154 GAS TUNGSTEN ARC WELDING PROCESS

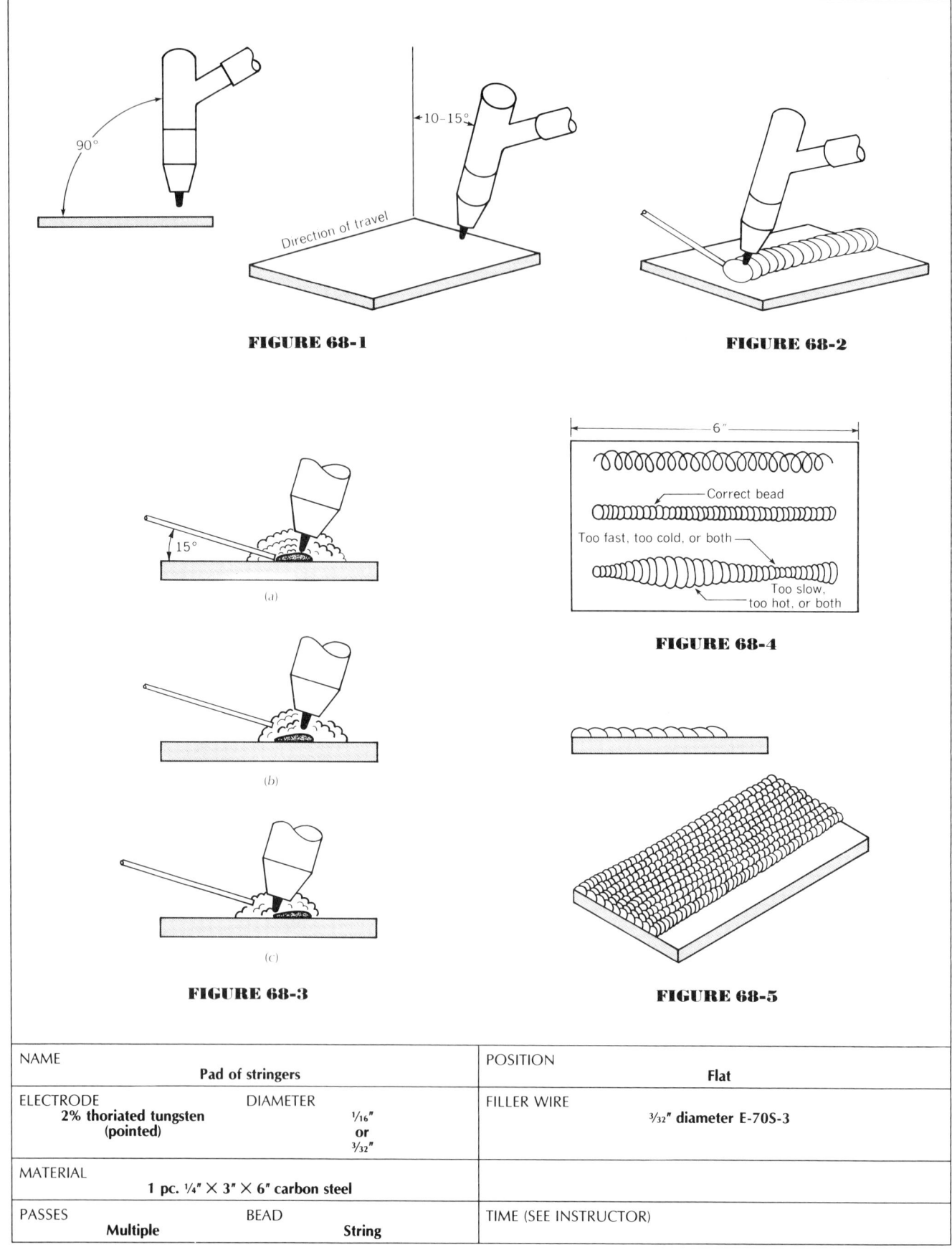

FIGURE 68-1

FIGURE 68-2

FIGURE 68-3

FIGURE 68-4

FIGURE 68-5

NAME Pad of stringers	POSITION Flat
ELECTRODE 2% thoriated tungsten (pointed) DIAMETER 1/16″ or 3/32″	FILLER WIRE 3/32″ diameter E-70S-3
MATERIAL 1 pc. ¼″ × 3″ × 6″ carbon steel	
PASSES Multiple BEAD String	TIME (SEE INSTRUCTOR)

Procedure 68

Running a Flat Pad of Stringers

OBJECTIVE

Upon completion of this lesson you should be able to weld a pad of stringers.

Text Reference: Section IV, Lesson 4A; page 308.

PROCEDURE

1. Adjust the welding current direct current electrode negative, 90 to 150 amperes. High frequency start 10 to 15 CFH argon, ½ in. gas cup.
2. Grind or sand the surface to be welded to remove any mill scale or rust.
3. Position the plate flat. Hold the GTAW torch in your right hand and position the torch just above the plate on the right-hand side of the plate. Grasp the filler metal between your thumb and forefinger in your left hand, allowing approximately 10 to 12 in. to protrude through your fingers.
4. Energize the torch. This can be done by turning on the power supply, turning on a torch switch or depressing a remote control foot switch. If the equipment is available use all three techniques to practice starting the arc. In future lessons we will just say strike the arc.
5. Slowly lower the torch and establish an arc. If you have high frequency it will cross the gap between the tungsten electrode and the workpiece. This will allow the current to begin to flow. When the arc is established the high frequency will stop. If you do not have high frequency you will have to touch the tungsten to the work lightly, then lift the tungsten off to establish an arc. This is best accomplished by holding the torch in such a manner that you can roll the torch in your hand to contact the workpiece and establish an arc.
6. Establish a weld puddle holding the torch perpendicular to the plate pointing toward the direction of travel. (See Figure 68-1.)
7. Bring the filler metal close to the puddle and add the wire to the front edge of the puddle. (See Figure 68-2.) Be careful not to touch the filler wire to the tungsten. This will cause the wire to melt onto the tungsten, contaminating it. If this happens you will have to stop and regrind your tungsten.
8. Add the filler metal a dab at a time to the leading edge of the puddle. Move the wire back and move the torch forward to advance the puddle continuing this technique to weld a straight bead. (See Figure 68-3.) Keep the end of the filler metal within the protective gas shield. If you withdraw the filler wire from the gas shield, the hot end will oxidize in the air and cause contaminants to enter the weld puddle.
9. Continue progressing at an even travel adding filler wire regularly to produce a straight bead with even width and height. Use a slight circular motion to spread the puddle and assure that the sides of the puddle fuse into the base metal. (See Figure 68-4.)
10. When you break the arc either to reposition yourself or because you have finished the bead, be sure to keep the torch over the puddle for a period of time to allow the postflow gas purge to protect the weld from oxidation while it is still hot.
11. Continue to weld stringer beads, overlapping the beads so that the bead your arc welding covers half of the previous bead. (See Figure 68-5.) Cool your plate is a quench tank regularly so it does not overheat. Be careful to avoid steam burns.

156 GAS TUNGSTEN ARC WELDING PROCESS

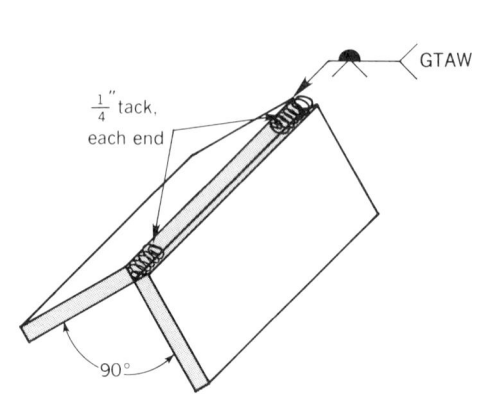

FIGURE 69-1

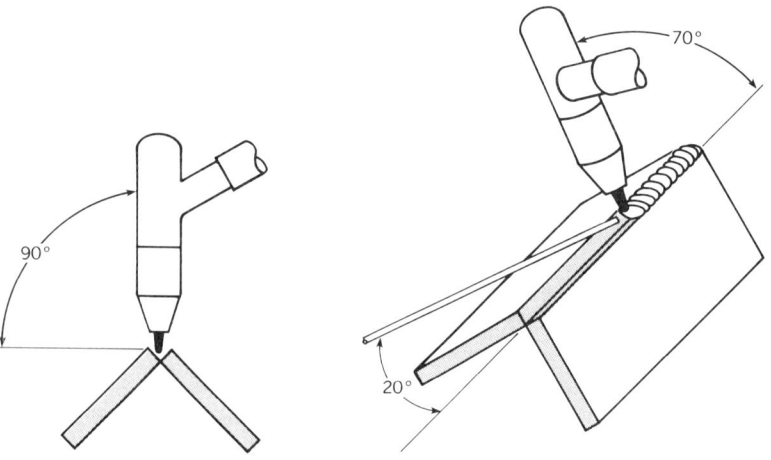

FIGURE 69-2

NAME Outside corner joint	POSITION Flat
ELECTRODE 2% thoriated tungsten (pointed) DIAMETER 3/32"	FILLER WIRE 1/16" or diameter E-70S-3 3/32"
MATERIAL 2 pcs. 1/8" × 1 1/2" × 6" carbon steel	
PASSES Single BEAD String	TIME (SEE INSTRUCTOR)

Procedure 69

Outside-Corner Joint, 1G Position

OBJECTIVE

Upon completion of this lesson you should be able to weld an outside-corner joint.

Text Reference: Section IV, Lesson 4A; page 309.

PROCEDURE

1. Adjust the welding current direct current electrode negative, 50 to 100 amperes. High frequency start, 10 to 15 CFH argon, ⅜ in. gas cup.
2. Grind or sand the surfaces to be welded. Tack the pieces together and position them as shown in Figure 69-1.
3. Beginning at the right side of the joint use the torch angles shown in Figure 69-2 and strike an arc. Establish a keyhole for complete penetration and begin adding filler wire to the leading edge of the puddle. Since the GTAW arc does not transfer metal, it is not too difficult to form the keyhole. If you use the correct amperage and add the filler metal properly, you will find you have a good deal of control over the weld puddle.
4. As you add the filler metal it will begin to shorten and your fingers holding the wire will begin to get close to the arc. When it becomes necessary to get another hold on your filler wire, break the arc keeping the torch with its postpurge gas flow over the cooling weld puddle, while keeping the end of the wire also in the gas purge. This protects the end of the wire from oxidation.
5. Extend the wire again 10 to 12 in., strike an arc, establish a puddle, and begin adding the wire to the lead edge of the puddle.
6. Remember to dab the wire into the puddle. Do not continuously feed it. Doing so can lead to porosity, lack of fusion, and slag entrapment.

158 GAS TUNGSTEN ARC WELDING PROCESS

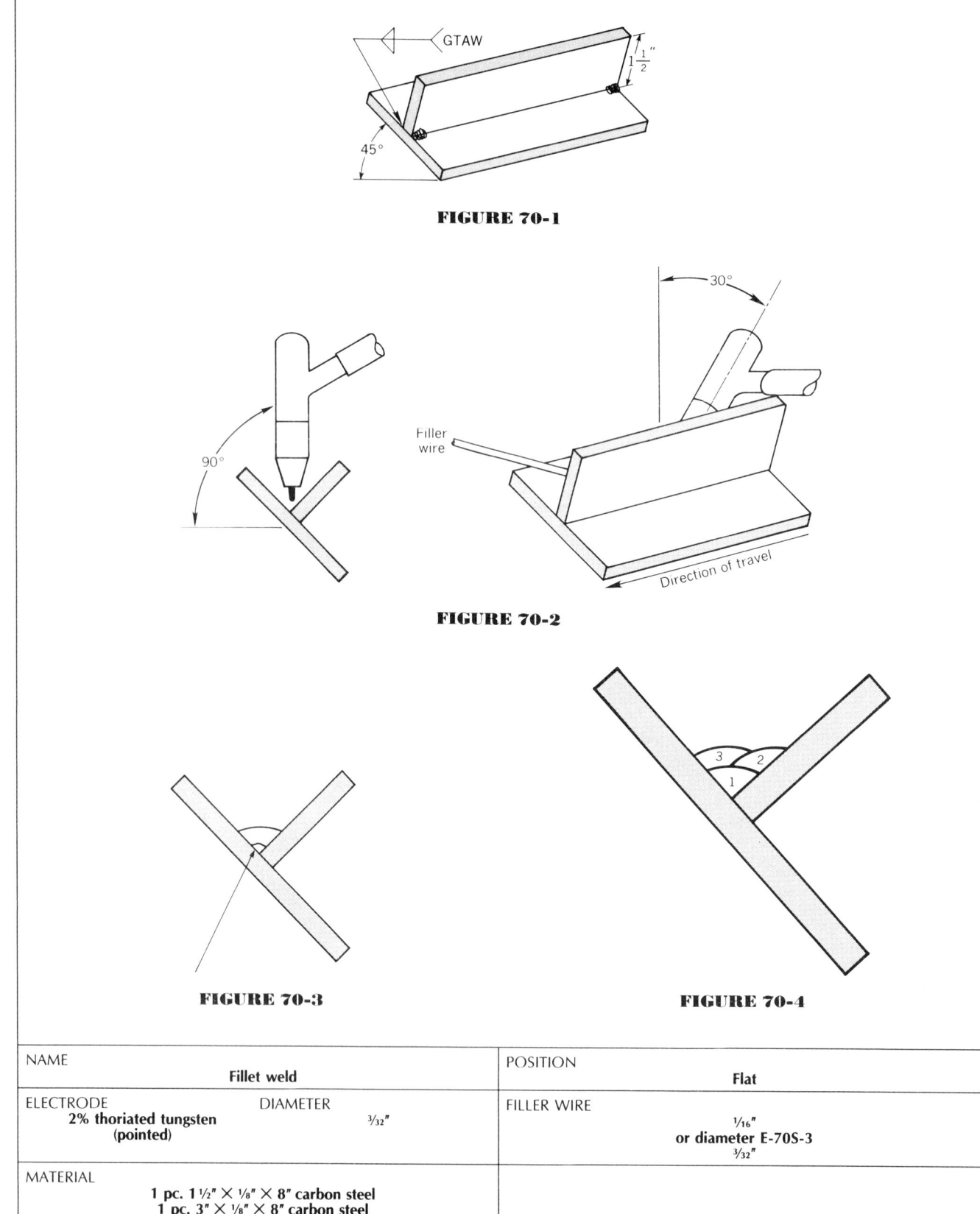

FIGURE 70-1

FIGURE 70-2

FIGURE 70-3

FIGURE 70-4

NAME			POSITION	
	Fillet weld			**Flat**
ELECTRODE		DIAMETER	FILLER WIRE	
2% thoriated tungsten (pointed)		$3/32''$		$1/16''$ or diameter E-70S-3 $3/32''$
MATERIAL	1 pc. 1½″ × ⅛″ × 8″ carbon steel 1 pc. 3″ × ⅛″ × 8″ carbon steel			
PASSES		BEAD	TIME (SEE INSTRUCTOR)	
Multiple		**String**		

Procedure 70

T-Joint Fillet, 1F Position

OBJECTIVE

Upon completion of this lesson you should be able to weld flat fillets.

Text Reference: Section IV, Lesson 4A; page 310.

PROCEDURE

1. Adjust the welding current direct current electrode negative, 50 to 100 amperes. High frequency start, 10 to 15 CFH argon, ⅜ in. gas cup.
2. Grind or sand the pieces to be welded. Tack them together and position them as shown in Figure 70-1. Position the T-base at a 45-degree angle so the weld will be applied flat.
3. Hold the torch perpendicular to the joint pointing toward the direction of travel about 30 degrees. (See Figure 70-2.)
4. Strike an arc and establish a puddle. Make sure the side walls melt down to the root of the T. Because the side walls are nearer to the electrode than the root of the joint, the arc will go to the side walls and cause them to melt before the root of the joint does. If you begin to add filler metal before the root of the joint is molten the weld will bridge the joint and not penetrate to the root. (See Figure 70-3.)
5. Add the filler wire in a dabbing motion, advancing the torch when you withdraw the filler metal. When you do this, you advance the puddle and prepare it for the addition of more filler metal.

Remember: When withdrawing the wire keep the end in the protective gas shield.

6. Complete the bead, cool the assembly and add two more passes as shown in Figure 70-4.
7. Reposition the T and weld the other side as you did the first using three beads.

160 GAS TUNGSTEN ARC WELDING PROCESS

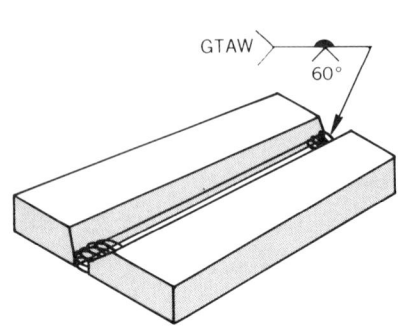

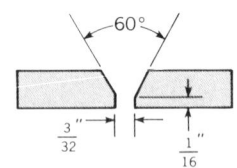

FIGURE 71-1

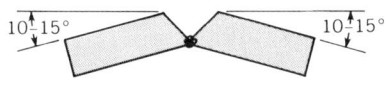

FIGURE 71-2

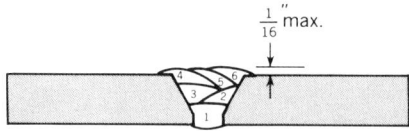

FIGURE 71-3

NAME Butt weld	POSITION Flat
ELECTRODE 2% thoriated tungsten (pointed) DIAMETER 3/32"	FILLER WIRE 1/16" or diameter E-70S-3 3/32"
MATERIAL 2 pcs. 1 1/4" × 3" × 6" carbon steel	
PASSES BEAD **Multiple** **String**	TIME (SEE INSTRUCTOR)

Procedure 71

V-Groove Joint, 1 G Position

OBJECTIVE

Upon completion of this lesson you should be able to weld butt joints in the flat position.

Text Reference: Section IV, Lesson 4A; page 311.

PROCEDURE

1. Prepare the ¼ in. steel plates by burning, machining, or grinding a 30-degree bevel on one 6 in. side of each plate. Grind a 1/16 in. root face on each piece.
2. Adjust the welding current direct current electrode negative, 90 to 150 amperes. High frequency start, 15 to 20 CFH Argon, ½ in. gas cup.
3. Tack the pieces together as shown in Figure 71-1 and position the plate so it is flat. After tacking, prebow the plates as shown in Figure 71-2 so the plate will be flat after welding.
4. Strike an arc and concentrate it on the root of the joint. Form a keyhole to ensure penetration and begin adding filler metal, by dabbing it to the leading edge of the puddle. Alternate the motion of adding the wire to the edge of the puddle, then work the puddle forward with the arc. Hold the torch perpendicular to the joint pointing toward the direction of travel approximately 20 degrees.

Remember: Every time you break the arc, hold the arc, hold the torch over the weld to provide shielding while the bead is cooling down.

5. Thoroughly clean the weld after each bead by wire brushing. Continue to weld the joint using the bead sequence shown in Figure 71-3.

Remember: The bead sequence shown is typical. Depending on the actual root spacing, groove angle, and rate of filler wire addition the number of passes may vary from those specified.

162 GAS TUNGSTEN ARC WELDING PROCESS

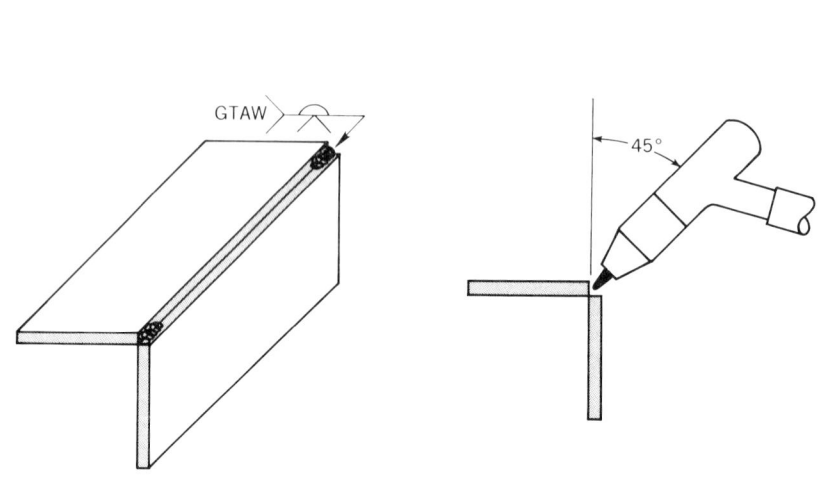

FIGURE 72-1

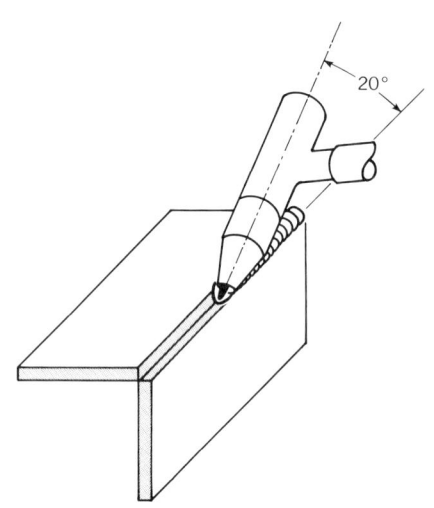

FIGURE 72-2

NAME	Corner joint	POSITION	Horizontal
ELECTRODE 2% thoriated tungsten (pointed)	DIAMETER 3/32"	FILLER WIRE	1/16" or diameter E-70S-3 3/32"
MATERIAL	2 pcs. 1/8" × 1 1/2" × 6" carbon steel		
PASSES Single	BEAD String	TIME (SEE INSTRUCTOR)	

Procedure 72

Outside-Corner Joint, 2G Position

OBJECTIVE

Upon completion of this lesson you should be able to weld corner joints in the horizontal position.

Text Reference: Section IV, Lesson 4B; page 312.

PROCEDURE

1. Adjust the welding current direct current electrode negative, 50 to 100 amperes. High frequency start, 10 to 15 CFH argon, ⅜ in. gas cup.
2. Grind or sand the pieces to be welded. Tack them together and position them as shown in Figure 72-1.
3. Begin at the right side of the joint and strike an arc. When an arc is established form a keyhole in the root of the joint. When the keyhole forms begin adding filler metal to the leading edge of the puddle.
4. Keep the torch perpendicular and point toward the direction of travel approximately 20 degrees. (See Figure 72-2.)
5. Complete the joint in one pass controlling the height of the bead by the rate at which you add the filler wire, and controlling the penetration by the force of the arc.

Note: The arc has some force and can be used to manipulate the weld puddle. Increasing the arc length will result in a softening of the arc, reducing the effect on the puddle.

6. Clean the joint thoroughly and inspect the bead for smoothness, and evenness of width. Check the root for penetration. If properly done, the root will show some weld along its entire length.

164 GAS TUNGSTEN ARC WELDING PROCESS

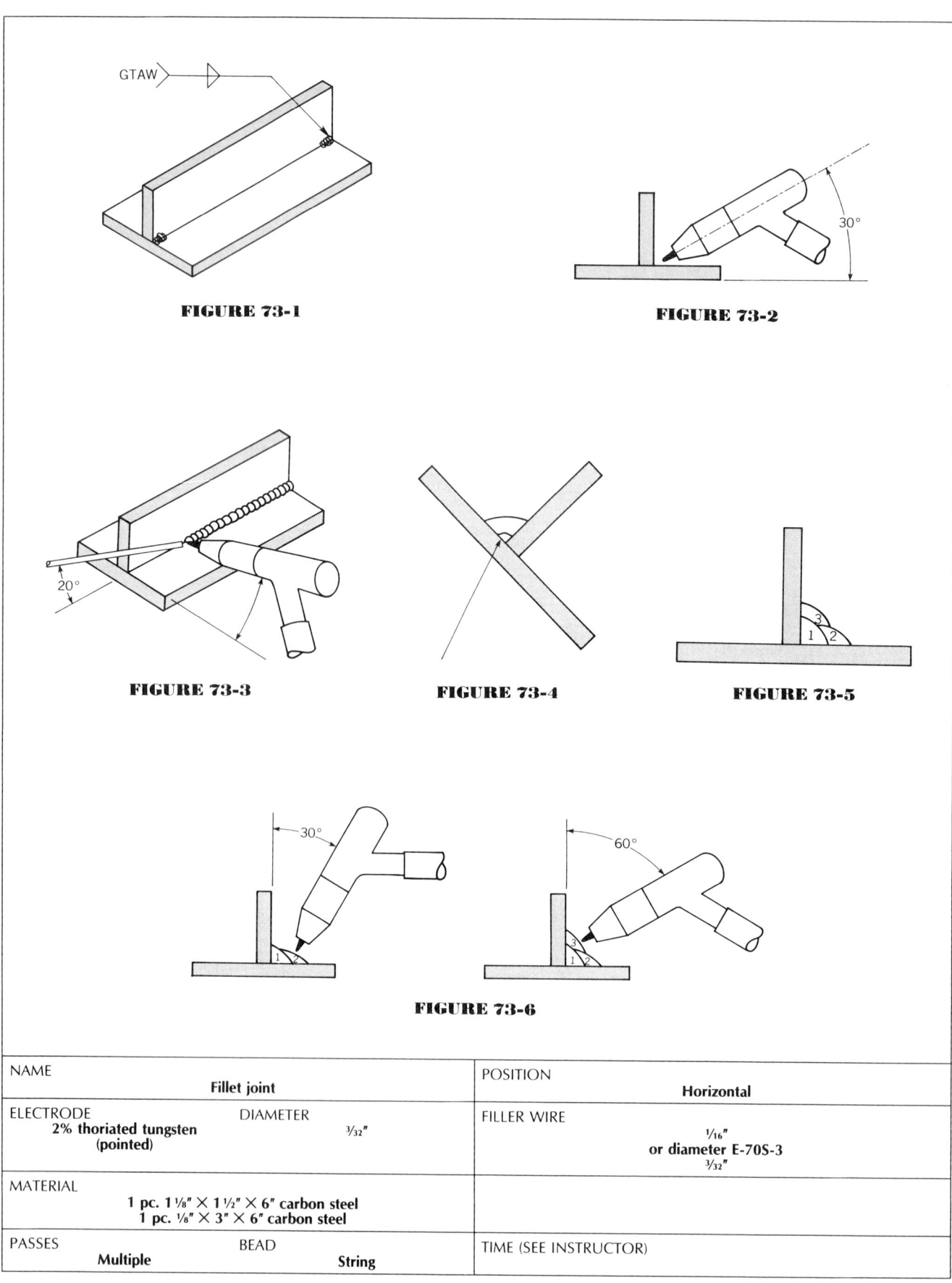

FIGURE 73-1

FIGURE 73-2

FIGURE 73-3

FIGURE 73-4

FIGURE 73-5

FIGURE 73-6

NAME	Fillet joint	POSITION	Horizontal
ELECTRODE 2% thoriated tungsten (pointed)	DIAMETER 3/32"	FILLER WIRE	1/16" or diameter E-70S-3 3/32"
MATERIAL	1 pc. 1 1/8" × 1 1/2" × 6" carbon steel 1 pc. 1/8" × 3" × 6" carbon steel		
PASSES Multiple	BEAD String	TIME (SEE INSTRUCTOR)	

Procedure 73

T-Joint Fillet, 2F Position

OBJECTIVE

Upon completion of this lesson you should be able to weld fillets in the horizontal position.

Text Reference: Section IV, Lesson 4B; page 313.

PROCEDURE

1. Adjust the welding current direct current electrode negative, 50 to 100 amperes. High frequency start, 10 to 15 CFH argon, ⅜ in. gas cup.
2. Grind or sand the pieces to be welded. Tack them together and position as shown in Figure 73-1.
3. Hold the torch as shown in Figure 73-2.

Note: The angle is changed to favor the vertical plate more than the flat plate. This will help to keep the vertical leg the same size as the flat leg by using the force of the arc to keep the metal up. Point the torch toward the direction of welding approximately 20 degrees.

4. Add the filler metal from the left side of the puddle at approximately a 20-degree angle from the flat. (See Figure 73-3.) Be careful as in the flat position to make sure you penetrate to the root of the joint. If you do not, you will get lack of penetration. (See Figure 73-4.)
5. When the first bead is completed, thoroughly clean it and deposit two more beads as shown in Figure 73-5. When welding bead 2 you will have to favor the bottom plate. Bead 3 will require you to adjust the torch angle to favor the vertical leg of the fillet. (See Figure 73-6.) Half of bead 2 should cover half of the first bead with the balance being deposited on the flat portion of the T. Half of bead 3 should cover the rest of bead 1 with the remainder being deposited on the vertical leg of the T.
6. When the first side is completed, cool the T and weld the second side.

166 GAS TUNGSTEN ARC WELDING PROCESS

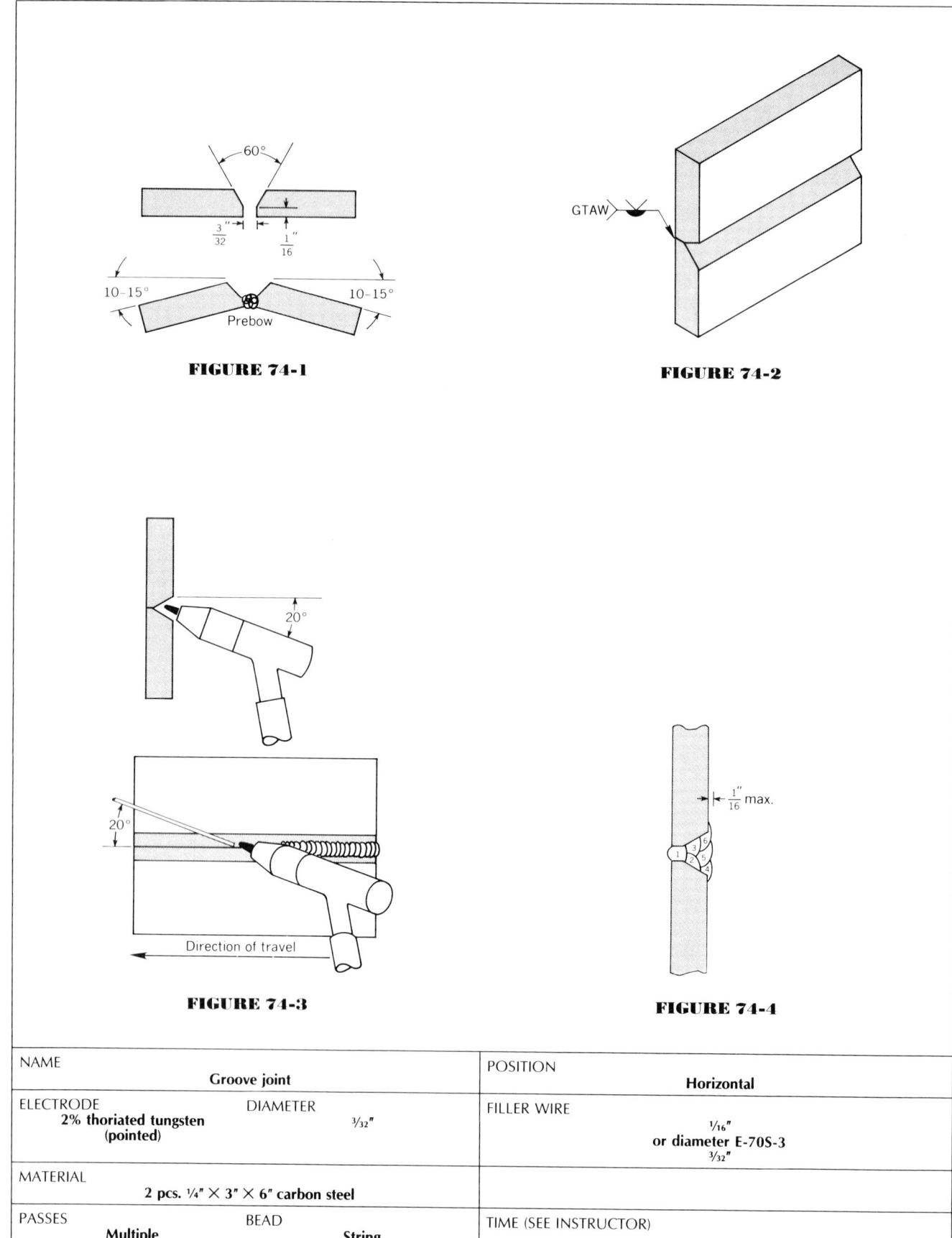

FIGURE 74-1

FIGURE 74-2

FIGURE 74-3

FIGURE 74-4

NAME	Groove joint	POSITION	Horizontal
ELECTRODE 2% thoriated tungsten (pointed)	DIAMETER 3/32"	FILLER WIRE	1/16" or diameter E-70S-3 3/32"
MATERIAL	2 pcs. 1/4" × 3" × 6" carbon steel		
PASSES Multiple	BEAD String	TIME (SEE INSTRUCTOR)	

Procedure 74

V-Groove Butt Joint, 2G Position

OBJECTIVE

Upon completion of this lesson you should be able to weld horizontal open-root V-groove butt joints.

Text Reference: Section IV, Lesson 4B; page 314.

PROCEDURE

1. Prepare one 6 in. side of each plate with a 30-degree bevel by burning, grinding, or machining. Grind a 1/16 in. root face on each piece.
2. Adjust the welding current direct current electrode negative, 90 to 150 amperes. High frequency start, 10 to 15 CFH argon, 1/2 in. gas cup.
3. Fit and tack the pieces, and prebow them for weld shrinkage as shown in Figure 74-1. Position the plate as shown in Figure 74-2.
4. Hold the torch so that it points toward the upper plate at about a 20-degree angle. Point the torch toward the direction of welding. (See Figure 74-3.) Strike an arc and establish a weld puddle. Watch the puddle until the root becomes molten. When this occurs begin adding the filler wire as shown in Figure 74-3 to the leading edge of the puddle. Use the torch angle and force of the arc to keep the puddle favoring the upper plate. Gravity will cause the puddle to flow to the lower plate.
5. Continue to weld the plate using the bead sequence shown in Figure 74-4.

Note: In the horizontal position the weld beads begin on the bottom plate and serve as a shelf for succeeding beads. Pay particular attention to interbead cleaning. On occasion, you will observe a smooth glass like deposit on the bead. This is silicon and should be removed before depositing another bead. If you do not you can trap this in the weld. If that occurs you have a discontinuity called slag entrapment.

6. Complete the joint, thoroughly clean it, cool it, and carefully examine it for complete and even root penetration. Check the face surface for contour, face reinforcement, and evenness of beads.

168 GAS TUNGSTEN ARC WELDING PROCESS

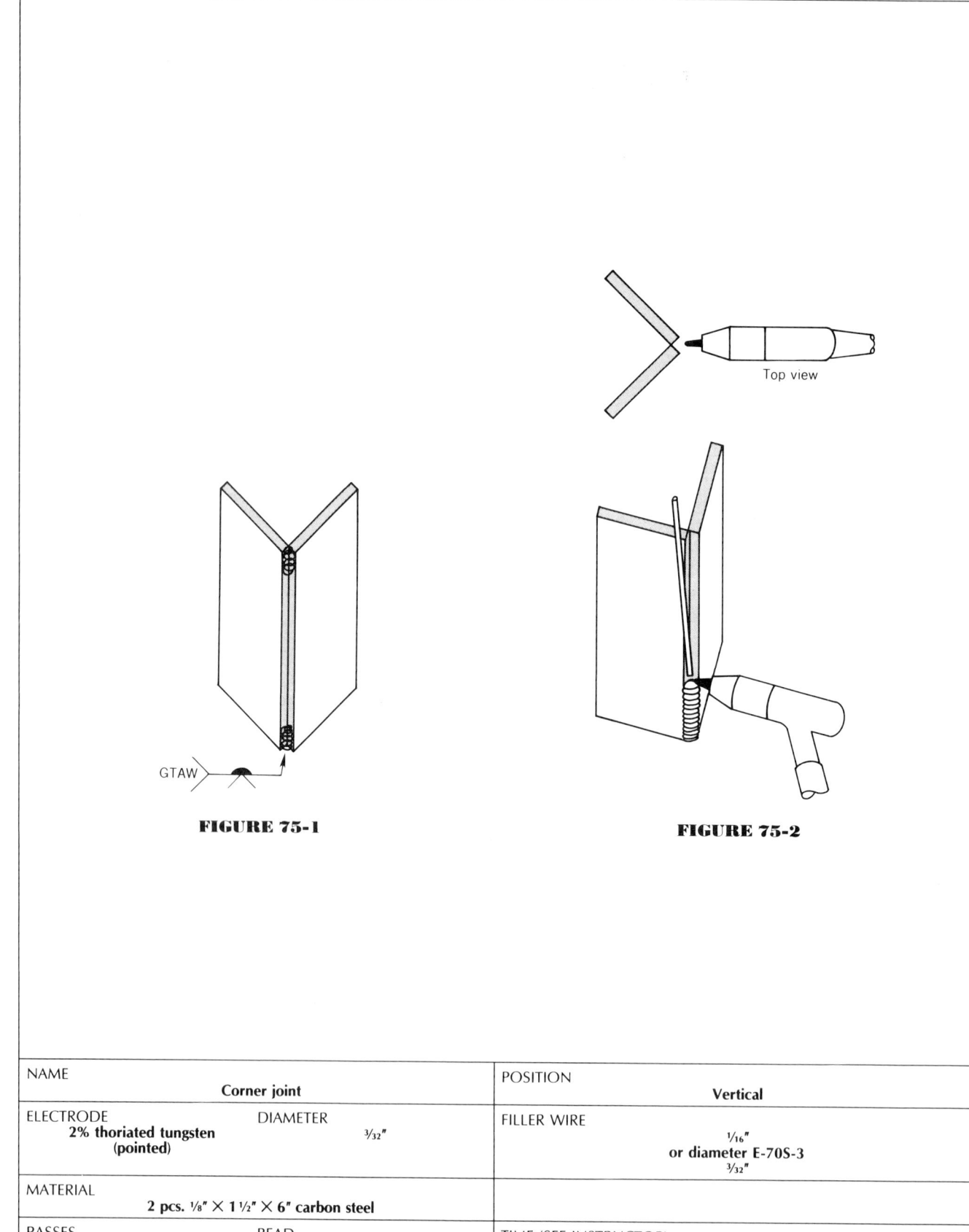

FIGURE 75-1

FIGURE 75-2

NAME	Corner joint	POSITION	Vertical
ELECTRODE 2% thoriated tungsten (pointed)	DIAMETER 3/32"	FILLER WIRE	1/16" or diameter E-70S-3 3/32"
MATERIAL	2 pcs. 1/8" × 1 1/2" × 6" carbon steel		
PASSES Single	BEAD String	TIME (SEE INSTRUCTOR)	

Procedure 75

Outside-Corner Joint, 3G Position

OBJECTIVE

Upon completion of this lesson you should be able to weld vertical corner joints.

Text Reference: Section IV, Lesson 4C; page 315.

PROCEDURE

1. Adjust the welding current direct current electrode negative, 50 to 100 amperes. High frequency start, 10 to 15 CFH argon, 3/8 in. gas cup.
2. Grind or sand the pieces to be welded. Tack them together and position them as shown in Figure 75-1.
3. Begin at the bottom of the joint and strike an arc. Hold the torch as shown in Figure 75-2, straight into the joint and pointing upward slightly in the direction of travel. Feed the wire from above into the leading edge of the puddle. (See Figure 75-2.)
4. Use a slight side-to-side weave and progress vertically up. Make sure you penetrate into the root of the joint. Build a shelf of weld at the bottom and use it to build on. Dab the filler wire into the leading edge of the puddle.
5. Complete the weld, cool it, clean it, and examine it for even root penetration and for evenness and smoothness of the face contour. The contour should be slightly convex.

170 GAS TUNGSTEN ARC WELDING PROCESS

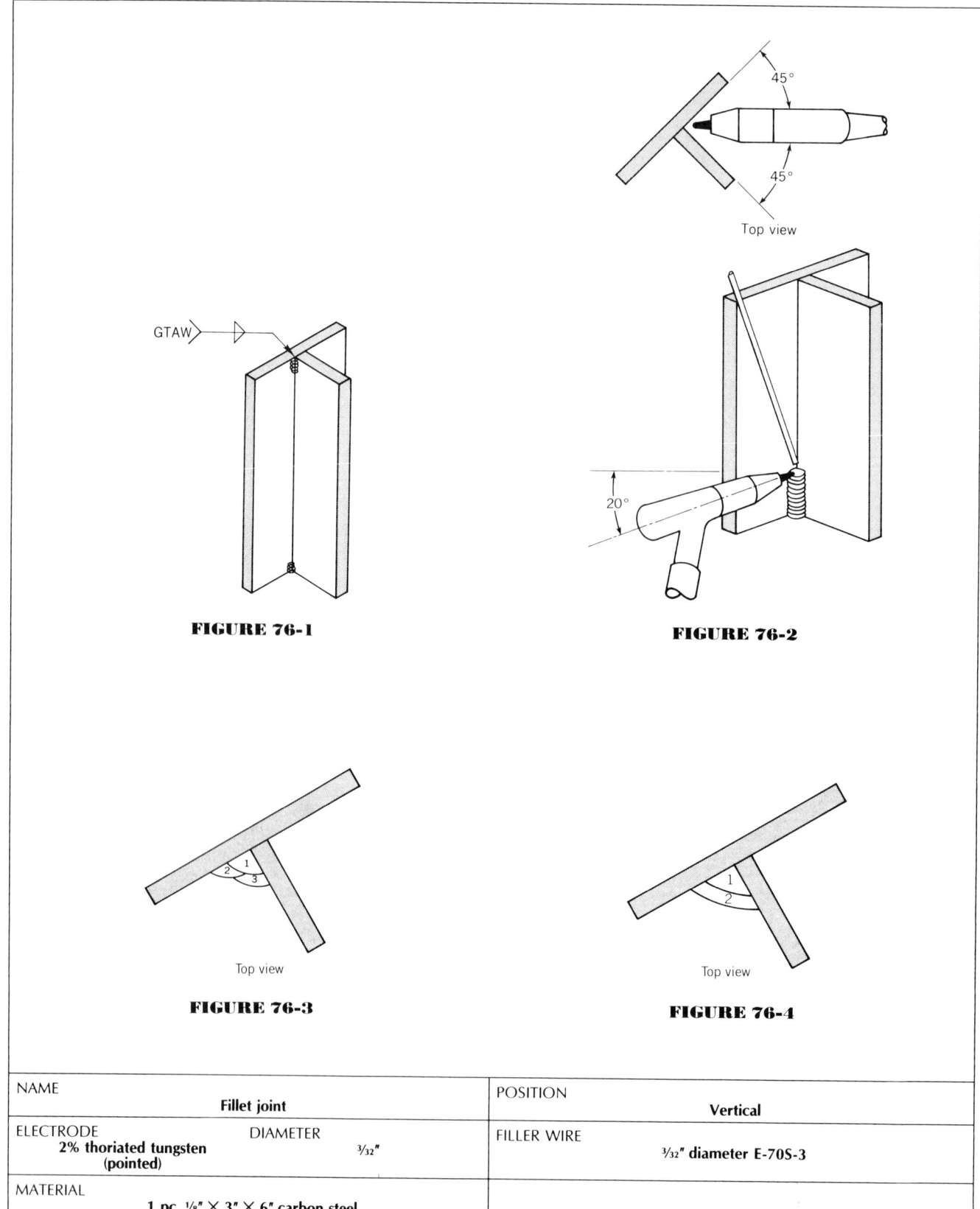

FIGURE 76-1

FIGURE 76-2

FIGURE 76-3

FIGURE 76-4

NAME	Fillet joint	POSITION	Vertical
ELECTRODE 2% thoriated tungsten (pointed)	DIAMETER 3/32"	FILLER WIRE	3/32" diameter E-70S-3
MATERIAL	1 pc. 1/8" × 3" × 6" carbon steel 1 pc. 1/8" × 1 1/2" × 6" carbon steel		
PASSES Multiple	BEAD String	TIME (SEE INSTRUCTOR)	

Procedure 76

T-Joint Fillet, 3F Position

OBJECTIVE

Upon completion of this lesson you should be able to weld vertical fillets.

Text Reference: Section IV, Lesson 4C; page 315.

PROCEDURE

1. Adjust the welding current direct current electrode negative, 50 to 100 amperes. High frequency start, 10 to 15 CFH Argon, ⅜ in. gas cup.
2. Grind or sand the pieces to be welded. Tack them together and position them as shown in Figure 76-1.
3. Using the torch angles shown in Figure 76-2, strike an arc at the bottom of the joint. Make sure the puddle extends into the root of the joint before you add filler wire. Failure to do so could result in your bridging the joint causing lack of penetration as we saw in previous lessons.
4. Establish a shelf to build your weld on. Use a slight side-to-side weave and add the filler wire from the top. (See Figure 76-2.)
5. After the first pass clean it thoroughly and apply a second layer consisting of two beads as shown in Figure 76-3.
6. After you have finished the three stringer bead side, cool the T and weld the second side using the weave-bead technique. Weld the first pass as a normal stringer bead. Put the second layer in using a weave bead, pausing at the sides to ensure the side walls are properly fused. Use the weave technique and bead sequence shown in Figure 76-4. When weaving a bead, the width of the weave should not exceed the size of the cup, in this case ⅜ in. The weld when finished should be slightly convex. The face should merge smoothly with the base metal at the toe of the weld.

172 GAS TUNGSTEN ARC WELDING PROCESS

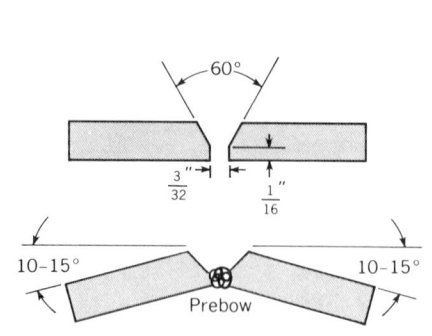

FIGURE 77-1

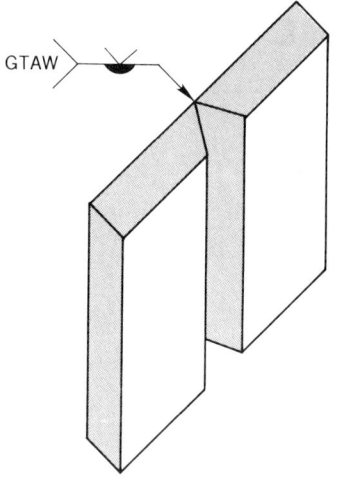

FIGURE 77-2

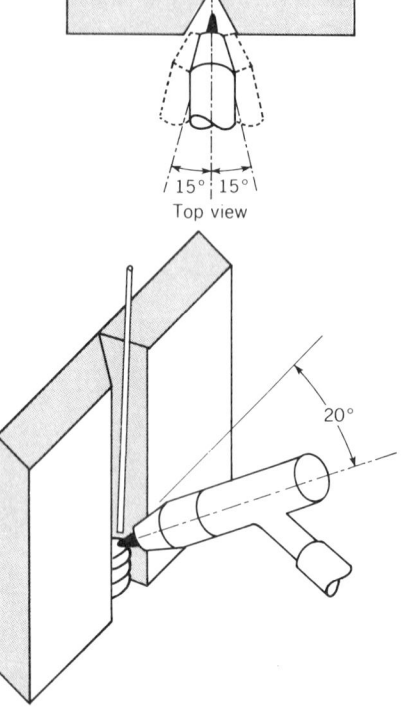

FIGURE 77-3

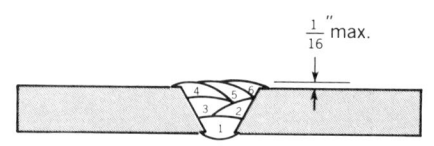

FIGURE 77-4

NAME	Groove joint	POSITION	Vertical
ELECTRODE 2% thoriated tungsten (pointed)	DIAMETER 3/32"	FILLER WIRE	1/16" or diameter E-70S-3 3/32"
MATERIAL	2 pcs. 1/4" × 3" × 6" carbon steel		
PASSES Multiple	BEAD String	TIME (SEE INSTRUCTOR)	

Procedure 77

V-Groove Butt Joint, 3G Position

OBJECTIVE

Upon completion of this lesson you should be able to weld vertical grooves.

Text Reference: Section IV, Lesson 4C; page 316.

PROCEDURE

1. Prepare one 6 in. side of each plate with a 30-degree bevel by burning, grinding, or machining. Grind a 1/16 in. root face on each piece.
2. Adjust the welding current, direct current electrode negative, 90 to 150 amperes. High frequency start, 10 to 15 CFH argon, 1/2 in. gas cup.
3. Fit and tack the pieces and prebow them for weld shrinkage, as shown in Figure 77-1. Put a 1/2 in. tack at each end. Position the plate as shown in Figure 77-2.
4. Strike an arc at the bottom of the joint. Melt the tack and make sure your puddle melts through the plate so a small keyhole appears. Use a slight side-to-side weave on the root to ensure root penetration and blending of the toe of the weld into the side walls of the groove. Use the torch angles shown in Figure 77-3.
5. Use a stringer-bead technique adding the filler metal as shown in Figure 77-3. Make sure you fill the crater completely at the top of the joint. Thoroughly clean the joint between beads being careful to remove any silicon deposit. Complete the joint using the stringer bead sequence shown in Figure 77-4.

174 GAS TUNGSTEN ARC WELDING PROCESS

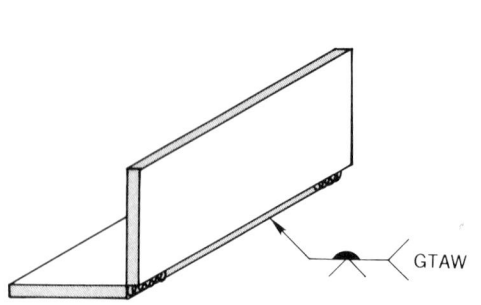

FIGURE 78-1

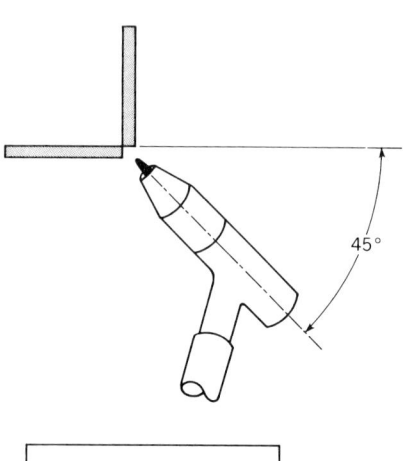

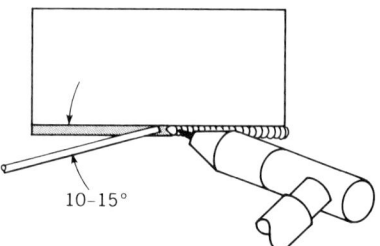

FIGURE 78-2

NAME	Corner joint	POSITION	Overhead
ELECTRODE 2% thoriated tungsten (pointed)	DIAMETER 3/32"	FILLER WIRE	1/16" or diameter E-70S-3 3/32"
MATERIAL 2 pcs. 1/8" × 1 1/2" × 6" carbon steel			
PASSES Single	BEAD String	TIME (SEE INSTRUCTOR)	

Procedure 78

Outside-Corner Joint, 4G Position

OBJECTIVE

Upon completion of this lesson you should be able to weld overhead-corner joints.

Text Reference: Section IV, Lesson 4D; page 318.

PROCEDURE

1. Adjust the welding current direct current electrode negative, 50 to 100 amperes. High frequency start, 10 to 15 CFH argon, ⅜ in. gas cup.
2. Grind or sand the pieces to be welded. Tack them together and position them as shown in Figure 78-1.
3. Hold the torch in the center of the joint pointing toward the direction of travel about 20 degrees. (See Figure 78-2.)
4. Strike an arc and establish a puddle making sure the puddle penetrates to the root of the joint. Begin adding filler metal as shown in Figure 78-2, to the leading edge of the puddle. Keep the travel speed and filler wire addition even to produce a bead with even width and build up.
5. Continue to the end of the joint. Cool the joint, clean it, and examine it for evenness of the width. Examine the face for smoothness and contour. The surface should be slightly convex.

176 GAS TUNGSTEN ARC WELDING PROCESS

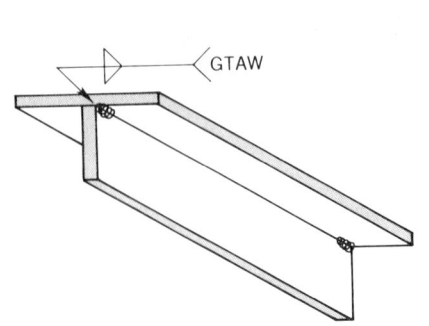

FIGURE 79-1

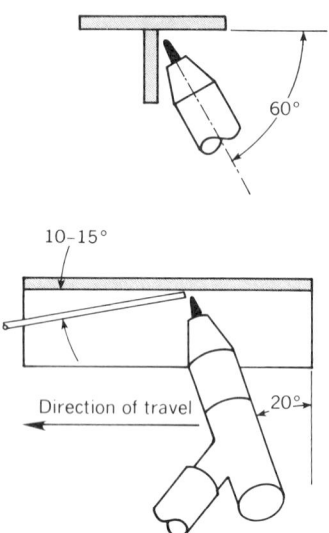

FIGURE 79-2

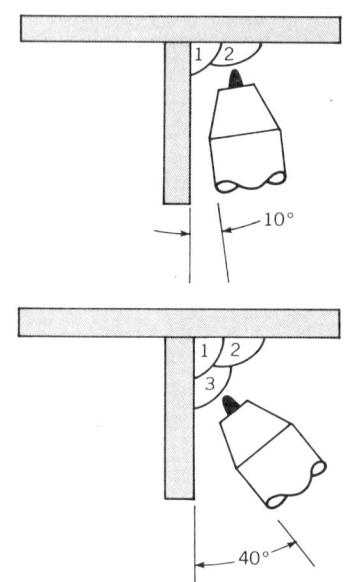

FIGURE 79-3

NAME	Fillet joint	POSITION	Overhead
ELECTRODE 2% thoriated tungsten (pointed)	DIAMETER 3/32"	FILLER WIRE	1/16" or diameter E-70S-3 3/32"
MATERIAL	1 pc. 1/8" × 1 1/2" × 6" carbon steel 1 pc. 1/8" × 3" × 6" carbon steel		
PASSES Multiple	BEAD String	TIME (SEE INSTRUCTOR)	

Procedure 79

T-Joint Fillet, 4F Position

OBJECTIVE

Upon completion of this lesson you should be able to weld overhead fillets.

Text Reference: Section IV, Lesson 4D; page 318.

PROCEDURE

1. Adjust the welding current, direct current electrode negative, 50 to 100 amperes. High frequency start, 10 to 15 CHF argon, ⅜ in. gas cup.
2. Grind or sand the pieces to be welded. Tack them together and position them as shown in Figure 79-1.
3. Find a comfortable position, supporting the torch and point it into the joint favoring the upper plate. Hold the torch at about 60 degrees from the plate and pointing toward the direction of travel approximately 20 degrees. (See Figure 79-2.)
4. Strike an arc and establish a puddle. Make sure the puddle reaches into the root of the joint. When you begin to add filler metal as shown in Figure 79-2, use the force of the arc to keep the metal on the overhead plate. Gravity will cause some metal to flow down on the vertical part of the T. By varying the angle of the torch you can control the leg length of the fillet.
5. Complete the first pass, cool the T assembly and thoroughly clean it. Add another layer consisting of two beads. You will have to change the torch angle for each bead. For bead two the torch will be concentrated on the overhead plate. For bead three the torch must be concentrated on the vertical plate. (See Figure 79-3.)
6. When you have completed the three beads, cool the T assembly, clean it, and weld the other side with a three bead fillet.

178 GAS TUNGSTEN ARC WELDING PROCESS

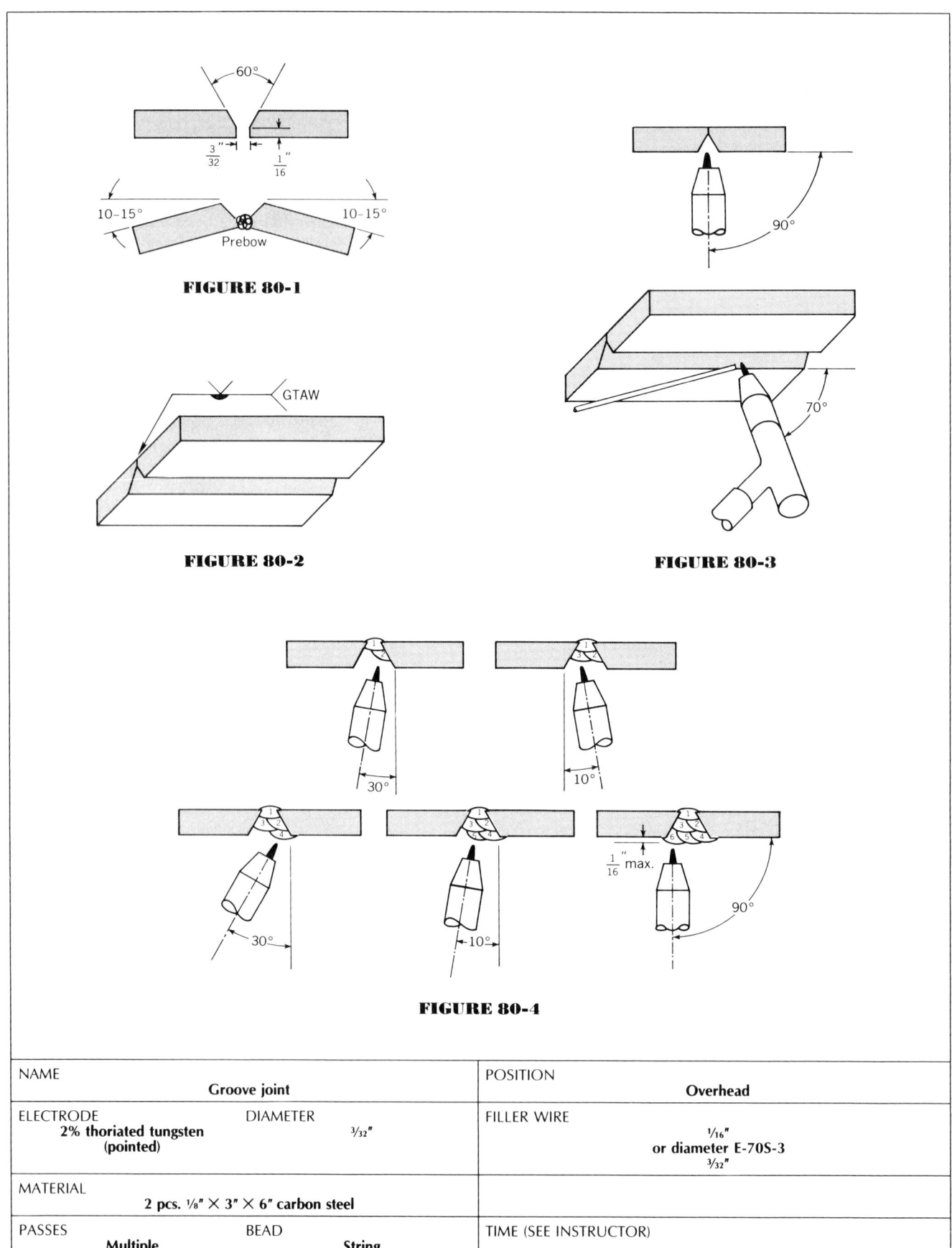

FIGURE 80-1

FIGURE 80-2

FIGURE 80-3

FIGURE 80-4

NAME			POSITION	
	Groove joint			Overhead
ELECTRODE		DIAMETER	FILLER WIRE	
2% thoriated tungsten (pointed)		3/32"		1/16" or diameter E-70S-3 3/32"
MATERIAL				
	2 pcs. 1/8" × 3" × 6" carbon steel			
PASSES		BEAD	TIME (SEE INSTRUCTOR)	
Multiple		String		

Procedure 80

V-Groove Butt Joint, 4G Position

OBJECTIVE

Upon completion of this lesson you should be able to weld overhead grooves.

Text Reference: Section IV, Lesson 4D; page 319.

PROCEDURE

1. Prepare one 6 in. side of each plate with a 30-degree bevel by burning, grinding, or machining. Grind a 1/16 in. root face on each piece.
2. Adjust the welding current, direct current electrode negative, 90 to 150 amperes. High frequency start, 15 to 20 CFH argon, 1/2 in. gas cup.
3. Fit and tack the pieces and prebow them for weld shrinkage, as shown in Figure 80-1. Put a 1/2 in. tack on each end. Position the plate as shown in Figure 80-2.
4. For the root pass, position the torch perpendicular to the joint, pointing toward the direction of travel. (See Figure 80-3.) As in the previous lessons, some support should be provided for the torch holding hand. This can consist of nothing more than extending a few fingers on that hand and having them rest on the plate you are about to weld on. Another method of providing support for the torch is to adjust the tungsten stick out, so that you can rest the cup on the plate. This will work when you are down in a groove but does not work when you are reinforcing the joint. Another drawback is that it makes it more difficult to see the weld puddle.
5. Strike an arc and establish a keyhole: add filler metal as shown in Figure 80-3. It may be necessary to weave the torch slightly to ensure good fusion between the toe of the weld and the side wall of the groove.
6. Add a second and third layer consisting of stringer beads as shown in Figure 8-4.

Note: The torch angle changes with each bead.

Gas Tungsten Arc Welding of Aluminum

In the following procedures you will apply to aluminum what you have learned about gas tungsten arc welding of carbon steel.

We will use the movements and motions we have learned and add some new skills to weld aluminum.

One of the major differences will be the type of current we will use. In the previous procedures, direct current electrode negative with a high frequency start was used to weld carbon steel. Aluminum is welded using alternating current with continuous high frequency. The continuous high frequency serves the purpose of creating a more stable arc as the current flow reverses direction.

GENERAL INSTRUCTIONS

These procedures will specify argon as the shielding gas. Helium may be used; however, higher gas flows are required, and the arc is not as stable as when using argon. Follow these general instructions for the procedures in this section:

1. Aluminum, to be properly welded, must be cleaned immediately prior to welding. The oxide formed on the surface of aluminum makes it extremely difficult to achieve a satisfactory weld. The surface can be cleaned chemically, or mechanically, by sanding, grinding, or wire brushing. If wire brushes are used, they should be made of stainless steel to prevent iron contamination.
2. When welding aluminum with the gas tungsten arc welding process use a pure tungsten electrode with a balled end. The end can be balled by inserting into the GTAW torch a blunt tungsten electrode.
3. With the power supply set on direct current electrode positive, turn on the current and gradually increase the amperage. A ball will begin to form on the end of the electrode.
4. When the ball is correctly formed, turn off the current and change over the power supply to alternating current for welding aluminum.

MATERIALS AND EQUIPMENT

Personal equipment called for in this section consists of:

1. Welding shield with appropriate filter lens
2. Safety glasses
3. Short cuff leather gloves
4. Appropriate welding clothing
5. Stainless steel wire brush
6. Pliers

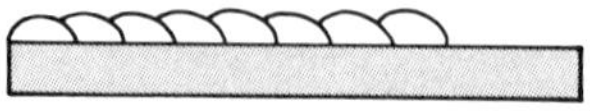

FIGURE 81-1

NAME Pad of stringers	POSITION Flat
ELECTRODE Pure tungsten (balled)　　DIAMETER 1/8"	FILLER WIRE 1/16" or diameter ER-4043 3/32"
MATERIAL 1 pc. 1/4" × 6" × 6" aluminum	
PASSES Multiple　　BEAD String	TIME (SEE INSTRUCTOR)

Procedure 81

Running a Flat Pad of Stringers

OBJECTIVE

Upon completion of this lesson you should be able to weld in the flat position.

Text Reference: Section IV, Lesson 5A; page 322.

PROCEDURE

1. Adjust the welding current, alternating (AC) continuous high frequency 175 to 210 amperes, 20 to 25 CFH argon, ½ in. gas cup.
2. Begin at the right side of the plate and strike an arc. The arc will have a buzzing sound, unlike the silence of a direct current arc.

Note: The area on the aluminum where the arc contacts the plate changes in appearance. This change is a result of the arc cleaning action. Regardless of how well you have cleaned the metal the arc will further clean the surface.

3. Aluminum will not show a lot of color as it melts because of its low melting point. When a puddle is developed it will appear very shiny.
4. When you have established the puddle, begin adding the filler metal to the leading edge of the puddle. After you have added filler metal withdraw it from the arc but keep it within the protective gas envelope. The hot filler wire will oxidize very rapidly if exposed to the atmosphere. Be careful not to hold the filler metal too close to the arc. Because of its low melting point, a large globule will form very quickly on the end of the wire if it is allowed to remain near the arc.
5. Run stringer beads the length of the plate overlapping them with the last weld covering one-half of the first. (See Figure 81-1.)
6. Be sure to fill the crater every time you break an arc. Hold the torch over the weld end for 15 to 20 seconds after you extinguish the arc to protect the hot weld from being exposed to the atmosphere.
7. Cool the plate frequently during welding to retain more control over the puddle. Examine every bead for even bead width, good bead contour, and smooth surface.

184 GAS TUNGSTEN ARC WELDING PROCESS

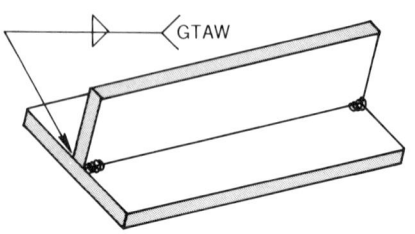

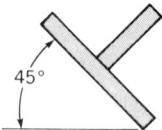

FIGURE 82-1

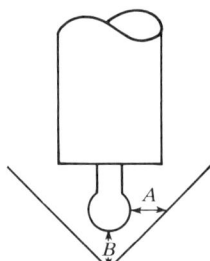

FIGURE 82-2

NAME			POSITION	
	Fillet			**Flat**
ELECTRODE		DIAMETER	FILLER WIRE	
Pure tungsten (balled)		³⁄₃₂″		¹⁄₁₆″ or diameter ER-4043 ³⁄₃₂″
MATERIAL				
	1 pc. ⅛″ × 1½″ × 6″ **aluminum**			
	1 pc. ⅛″ × 3″ × 6″ **aluminum**			
PASSES		BEAD	TIME (SEE INSTRUCTOR)	
	Multiple	**String**		

Procedure 82

T-Joint Fillet, 1F Position

OBJECTIVE

Upon completion of this lesson you should be able to weld flat fillets.

Text Reference: Section IV, Lesson 5A; page 322.

PROCEDURE

1. Adjust the welding current, alternating (AC) continuous high frequency, 125 to 160 amperes, 20 to 25 CFH argon, ⅜ in. gas cup.
2. Clean the pieces, tack them together, and position them as shown in Figure 82-1. It may be necessary to add filler metal to make the tack.
3. Strike an arc on the right side of the T assembly and establish a puddle. Make sure the joint becomes molten to the root of the joint. This will not be as easy as it was when using direct current. The concentrated arc of direct current allows you to direct the heat of the arc exactly where you want it. Alternating current will radiate from the surface of the ball on the end of the tungsten and will arc to the closest surface. (See Figure 82-2.)

Note: Distance A is closer to the electrode than distance B, therefore the arc will be drawn across the shortest distance and will establish a puddle on the side wall at point A before the root (point B) melts.

4. Begin adding the filler wire to the leading edge of the puddle. Be careful not to get the wire too close to the electrode or it will melt onto the tungsten. This contaminates the tungsten and will require it to be dressed before further use.
5. Continue to the end of the joint applying a single pass fillet, cool the plate, and put a single pass fillet weld on the other side.
6. Cool the plate and examine it for bead appearance, leg length, and undercut at the toe of the weld.

186 GAS TUNGSTEN ARC WELDING PROCESS

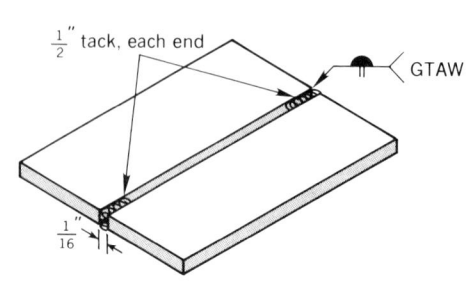

FIGURE 83-1

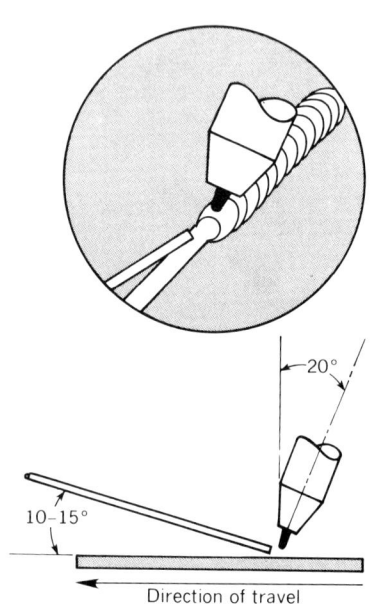

FIGURE 83-2

NAME Groove	POSITION Flat
ELECTRODE Pure tungsten (balled) DIAMETER 1/8"	FILLER WIRE 1/16" or diameter ER-4043 3/32"
MATERIAL 2 pcs. 1/8" × 3" × 6" aluminum	
PASSES Multiple BEAD String	TIME (SEE INSTRUCTOR)

Procedure 83

Square-Groove Butt Joint, 1G Position

OBJECTIVE

Upon completion of this lesson you should be able to weld flat butts.

Text Reference: Section IV, Lesson 5A; page 323.

PROCEDURE

1. Adjust the welding current, alternating (AC) continuous high frequency 175 to 210 Amperes, 20 to 25 CFH argon, ½ in. gas cup.
2. Clean the pieces, tack them together, and position them as shown in Figure 83-1.
3. Strike an arc and establish a keyhole. When the keyhole is formed begin adding filler wire to the leading edge of the puddle. (See Figure 83-2.) Pay attention to the penetration and bead width.
4. Progress from the right to the left. As you progress the plate will heat up causing the puddle to get larger. You can control this by either decreasing the amperage or increasing your travel speed.
5. Complete the joint, cool it and examine it for face appearance and toe undercut. The penetration through the joint should be complete with even width and root reinforcement.

188 GAS TUNGSTEN ARC WELDING PROCESS

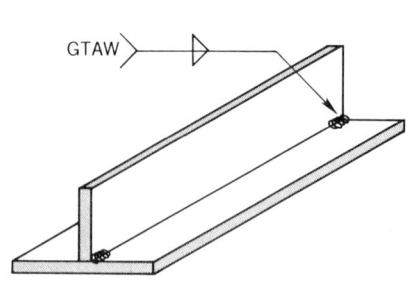

FIGURE 84-1

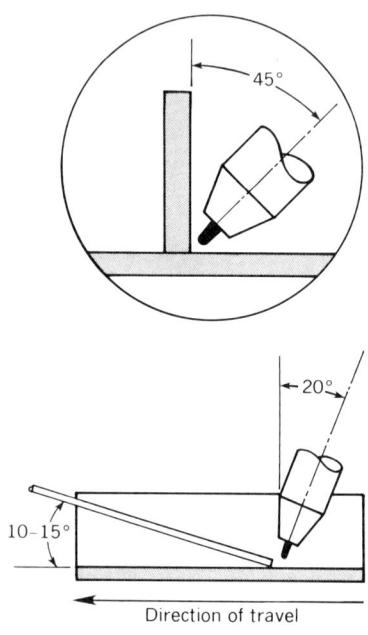

FIGURE 84-2

NAME Fillet	POSITION Horizontal
ELECTRODE **Pure tungsten (balled)** DIAMETER 3/32"	FILLER WIRE 1/16" or diameter ER-4043 3/32"
MATERIAL 1 pc. 1/8" × 1 1/2" × 6" aluminum 1 pc. 1/8" × 3" × 6" aluminum	
PASSES **Multiple** BEAD **String**	TIME (SEE INSTRUCTOR)

Procedure 84

T-Joint Fillet, 2F Position

OBJECTIVE

Upon completion of this lesson you should be able to weld horizontal fillets.

Text Reference: Section IV, Lesson 5B; page 324.

PROCEDURE

1. Adjust the welding current, alternating (AC) continuous high frequency, 125 to 160 amperes, 20 to 25 CFH argon, ⅜ in. gas cup.
2. Clean the pieces, tack them together and position them as shown in Figure 84-1.

Remember: It may be necessary to use filler wire to make the tack.

3. Strike an arc on the right side of the T assembly and establish a puddle. Use a torch angle of 45 degrees to the joint, with the torch pointing toward the direction of travel approximately 20 degrees. (See Figure 84-2.)
4. Concentrate the arc evenly on both the vertical and horizontal legs so you apply an equal leg fillet weld.
5. Complete the weld, cool the joint, and weld the other side, cool the assembly and examine the weld for even face contour, equal leg length, and lack of undercut at the toe of the weld.

190 GAS TUNGSTEN ARC WELDING PROCESS

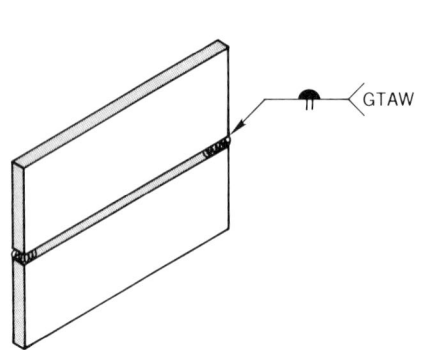

FIGURE 85-1

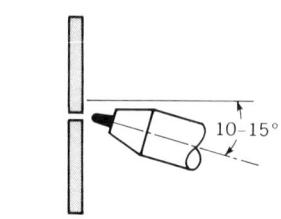

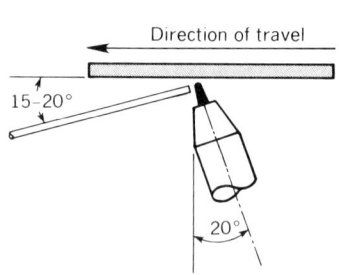

FIGURE 85-2

FIGURE 85-3

NAME Groove	POSITION Horizontal
ELECTRODE Pure tungsten (balled) DIAMETER 1/8"	FILLER WIRE 1/16" or diameter ER-4043 3/32"
MATERIAL 2 pcs. 1/8" × 3" × 6" aluminum	
PASSES Multiple BEAD String	TIME (SEE INSTRUCTOR)

Procedure 85

Square-Groove Butt Joint, 2G Position

OBJECTIVE

Upon completion of this lesson you should be able to weld horizontal grooves.

Text Reference: Section IV, Lesson 5B; page 324.

PROCEDURE

1. Adjust the welding current, alternating (AC) continuous high frequency, 175 to 210 amperes, 20 to 25 CFH argon, ½ in. gas cup.
2. Clean the pieces, tack them together and position them as shown in Figure 85-1.
3. Strike an arc and establish a puddle. Use the torch angle shown in Figure 85-2, positioning the torch so you favor the upper plate. Point toward the direction of travel approximately 20 degrees.
4. Establish a keyhole and add filler metal to the leading edge of the puddle at an angle of 15 to 20 degrees. (See Figure 85-2.)

Remember: Keep the end of the wire within the protective gas envelope. Pay attention to the keyhole and watch for even root penetration. You may find it necessary to use a slight back and forth motion of the torch to keep root penetration consistent. (See Figure 85-3.)

5. Cool the plate, clean it, and examine it for even root penetration, bead width, and face reinforcement.

192 GAS TUNGSTEN ARC WELDING PROCESS

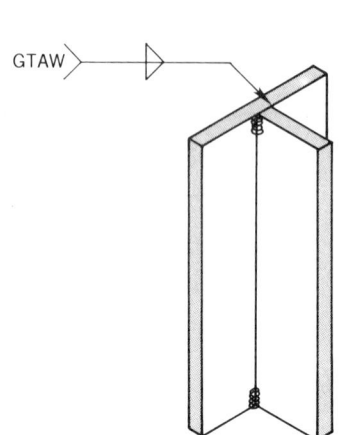

FIGURE 86-1

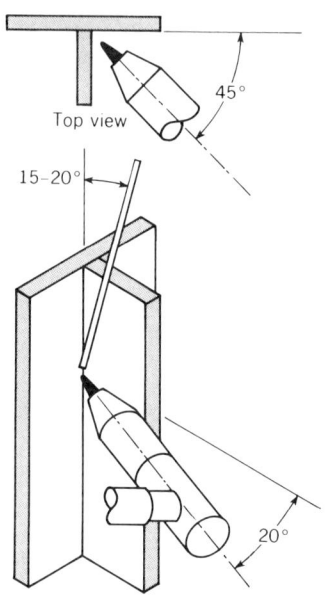

FIGURE 86-2

NAME	Fillet	POSITION	Vertical
ELECTRODE Pure tungsten (balled)	DIAMETER 3/32"	FILLER WIRE	1/16" or diameter ER-4043 3/32"
MATERIAL	1 pc. 1/8" × 1 1/2" × 6" aluminum 1 pc. 1/8" × 3" × 6" aluminum		
PASSES Multiple	BEAD String	TIME (SEE INSTRUCTOR)	

Procedure 86

T-Joint Fillet, 3F Position

OBJECTIVE

Upon completion of this lesson you should be able to weld vertical fillets.

Text Reference: Section IV, Lesson 5C; page 325.

PROCEDURE

1. Adjust the welding current, alternating (AC) continuous high frequency, 125 to 160 amperes, 20 to 25 CFH argon, 3/8 in. gas cup.
2. Clean the pieces, tack them together and position them as shown in Figure 86-1.

Remember: It may be necessary to use filler wire to make the tack.

3. Strike an arc and establish a puddle. The sides will melt before the root. Make sure the side walls are melted to the root before you add filler metal or you will bridge the joint and get incomplete joint penetration. Use a torch angle of 45 degrees to the joint with the torch pointing upward approximately 20 degrees. (See Figure 86-2.)
4. Add the filler wire from above at a 15 to 20 degree angle from vertical. (See Figure 86-2.) Keep the torch steady without a weave and progress evenly vertically up. Dab the filler metal into the leading edge of the puddle. By dabbing the wire regularly and keeping your travel speed even you will produce beads with even size and good contour.
5. When you have finished the first side, cool the T bar, clean it, examine the weld, then weld the second side.

194 GAS TUNGSTEN ARC WELDING PROCESS

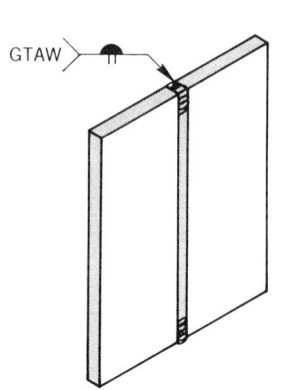

FIGURE 87-1

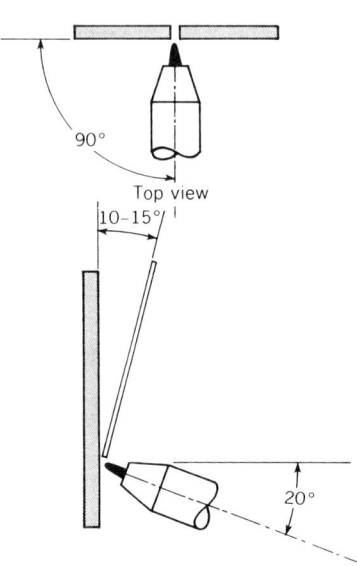

FIGURE 87-2

NAME			POSITION	
	Groove			Vertical
ELECTRODE		DIAMETER	FILLER WIRE	
Pure tungsten (balled)		1/8"		1/16" or diameter ER-4043 3/32"
MATERIAL				
	2 pcs. 1/8" × 3" × 6" aluminum			
PASSES		BEAD	TIME (SEE INSTRUCTOR)	
Multiple		String		

Procedure 87

Square-Groove Butt Joint, 3G Position

OBJECTIVE

Upon completion of this lesson you should be able to weld vertical groove joints.

Text Reference: Section IV, Lesson 5C; page 326.

PROCEDURE

1. Adjust the welding current, alternating (AC) continuous high frequency 175 to 210 amperes, 20 to 25 CFH argon, ½ in. gas cup.
2. Clean the pieces, tack them together and position them as shown in Figure 87-1.
3. Strike an arc and establish a puddle. Use the torch angle as shown in Figure 87-2. Point toward the direction of travel approximately 20 degrees.
4. Look for the keyhole before adding wire. When the keyhole forms add the filler wire from the top at a 15- to 20-degree angle. Dab it in the leading edge of the puddle evenly. This along with an even travel speed will produce an even bead with good penetration.
5. Add enough filler wire to reinforce the face of the weld to a maximum of 1/16 in. Root reinforcement should also not exceed 1/16 in.
6. Complete the joint, cool it, clean it, and examine the face and root surfaces for appearance and size.

196 GAS TUNGSTEN ARC WELDING PROCESS

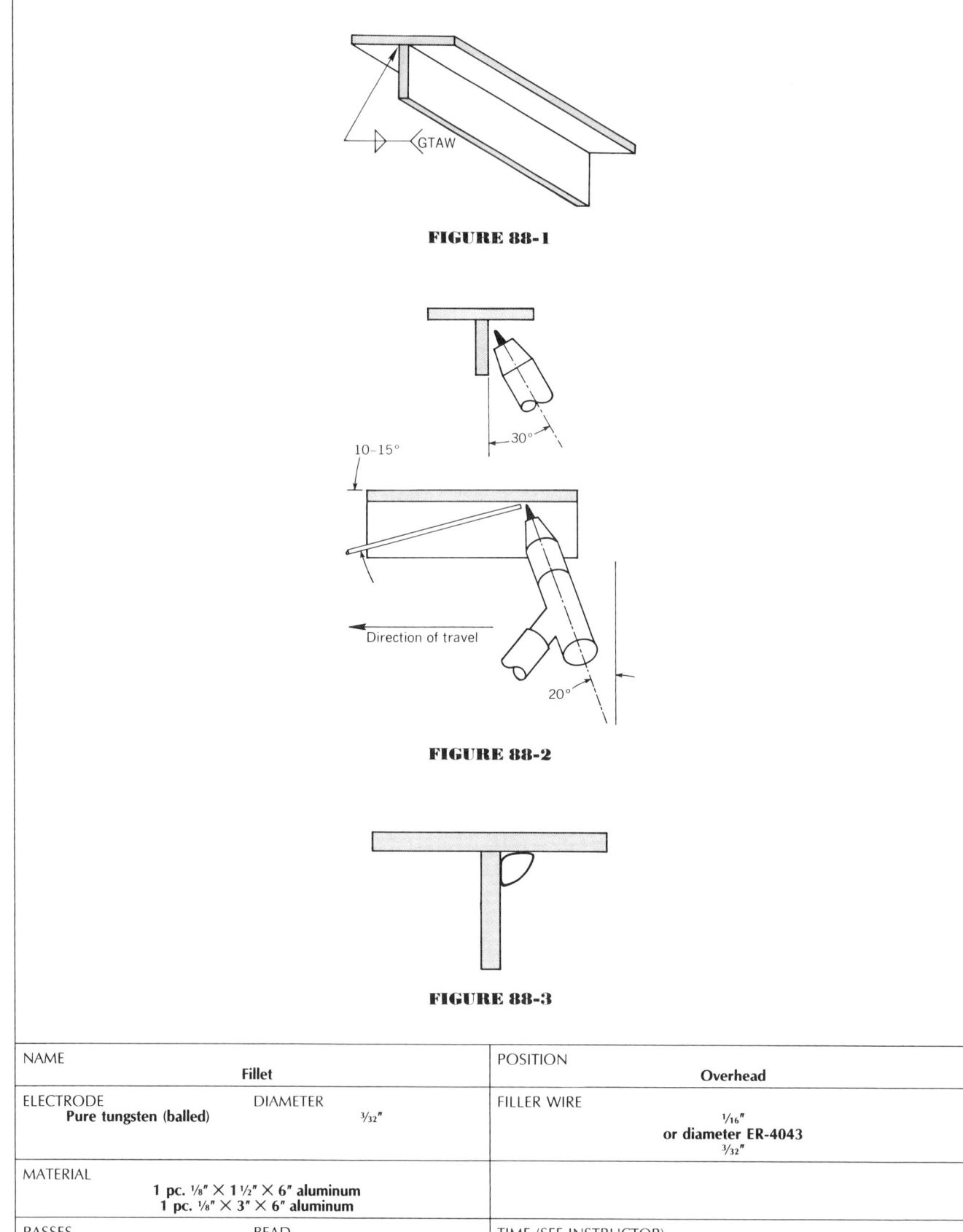

FIGURE 88-1

FIGURE 88-2

FIGURE 88-3

NAME			POSITION	
	Fillet			**Overhead**
ELECTRODE		DIAMETER	FILLER WIRE	
Pure tungsten (balled)		3/32"		1/16" or diameter ER-4043 3/32"
MATERIAL				
	1 pc. 1/8" × 1 1/2" × 6" aluminum			
	1 pc. 1/8" × 3" × 6" aluminum			
PASSES		BEAD	TIME (SEE INSTRUCTOR)	
Multiple		**String**		

Procedure 88

T-Joint Fillet, 4F Position

OBJECTIVE

Upon completion of this lesson you should be able to weld overhead fillets.

Text Reference: Section IV, Lesson 5D; page 327.

PROCEDURE

1. Adjust the welding current, alternating (AC) continuous high frequency, 125 to 160 amperes, 10 to 25 CFH argon, ⅜ in. gas cup.
2. Clean the pieces, tack them together and position them as shown in Figure 88-1.
3. Strike an arc on the right side of the joint. Establish a puddle, making sure the joint is molten to the root before adding filler metal. Position the torch to favor the top plate. (See Figure 88-2.) Gravity will help the puddle to flow down on the vertical plate to form an equal leg fillet.
4. Add filler metal to the front edge of the puddle as shown in Figure 88-2.

Remember: Dab the metal into the puddle. Continuous addition of the filler metal could lead to lack of fusion at the root of the joint. (See Figure 88-3.) This results because the filler metal bridges the joint and does not allow the arc to penetrate to the root.

5. Continue to the end of the joint, applying a single pass fillet. Cool the plate and put a single pass fillet weld on the other side.

GAS TUNGSTEN ARC WELDING PROCESS

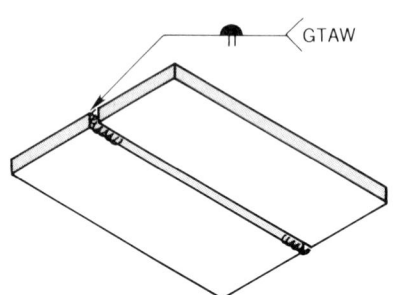

FIGURE 89-1

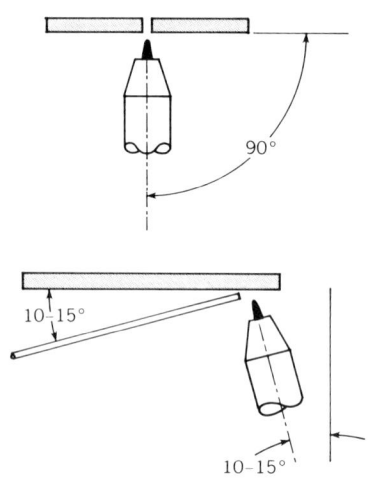

FIGURE 89-2

NAME		POSITION	
	Groove		Overhead
ELECTRODE	DIAMETER	FILLER WIRE	
Pure tungsten (balled)	1/8"		1/16" or diameter ER-4043 3/32"
MATERIAL			
2 pcs. 1/8" × 3" × 6" aluminum			
PASSES	BEAD	TIME (SEE INSTRUCTOR)	
Multiple	String		

Procedure 89

Square-Root Butt Joint, 4G Position

OBJECTIVE

Upon completion of this lesson you should be able to weld overhead grooves.

Text Reference: Section IV, Lesson 5D; page 328.

PROCEDURE

1. Adjust the welding current, alternating (AC) continuous high frequency, 175 to 210 amperes, 20 to 25 CFH argon, ½ in. gas cup.
2. Clean the pieces, tack them together and position them as shown in Figure 89-1.
3. Strike an arc and establish a keyhole. Hold the torch perpendicular to the joint and pointing toward the direction of travel 10 to 15 degrees. (See Figure 89-2.)
4. When the keyhole is established and you have penetration through the root of the joint begin adding filler wire to the leading edge of the puddle. (See Figure 89-2.) Add it using the dab method, regularly to the leading edge of the puddle. This will result in a bead of even width and reinforcement.
5. Complete the joint, cool it and examine it for evenness of bead width and height of reinforcement. Reinforcement should not exceed ¹⁄₁₆ in.

GAS METAL ARC WELDING PROCESS

Short-Circuiting Arc Welding of Steel

In the following lessons you will be welding carbon steel with the gas metal arc welding (GMAW) process, using the short circuiting method of transfer.

GENERAL INSTRUCTIONS

Review the following instructions before beginning the Procedures in this section.

1. Before you start to weld you must check out and set up your equipment. Inspect the equipment. If there is any equipment that needs repair notify your instructor.
2. Go to the power supply and check the cable connections. The correct current for GMAW short-circuiting transfer is DCEP.
3. Check the work cable and turn on the power supply.
4. Proceed to the wire feeder and check for the proper size filler wire and wire feed rolls. For the lessons in this section, you will be told to use .035 in. diameter E-70S-3 filler wire. If the wire is not threaded through the gun and cable assembly use the wire feeder switch or gun trigger to feed wire through the conduit assembly and through the gun.

Caution: If you use the gun trigger to feed the wire make sure the gun is not touching or near objects on which an arc might be struck. Remove the tip when the wire is being fed through the gun. When the wire is fed through release the wire feed button and install the contact tip. For short-circuiting transfer, the tip should protrude beyond the gas nozzle 1/8 in.

5. Set the gas flow to 25 to 35 CFH of 75% argon 25% carbon dioxide. Apply antispatter compound to the gas cup and contact tip to prevent spatter buildup.

MATERIALS AND EQUIPMENT

For this section the materials and equipment will be as follows:

1. Welding shield with appropriate lens
2. Safety glasses
3. Gauntlet type leather gloves
4. Appropriate welding clothing
5. Wire brush
6. Diagonal side cutters

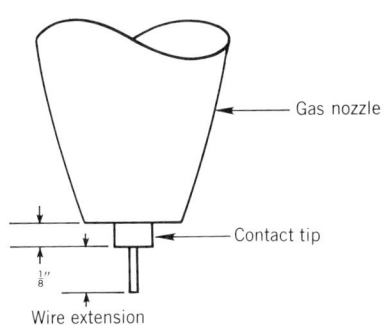

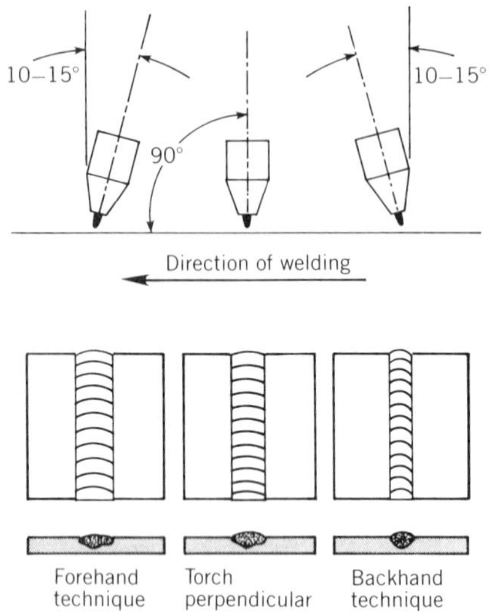

204 GAS METAL ARC WELDING PROCESS

FIGURE 90-1

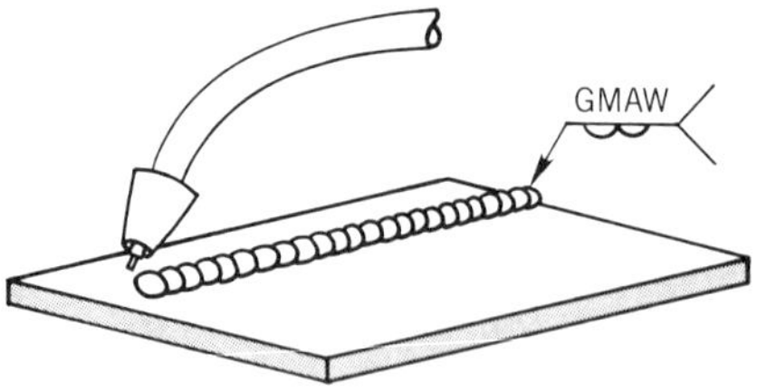

FIGURE 90-2

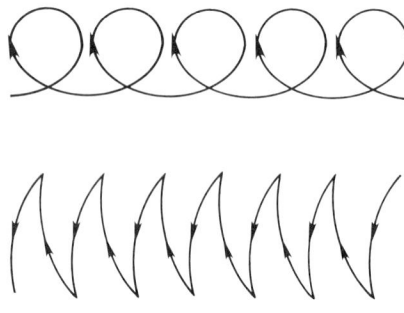

FIGURE 90-3

FIGURE 90-4

NAME		POSITION	
	Pad of stringer beads		Flat
ELECTRODE	DIAMETER	AMPERAGE	VOLTAGE
E-70S-3	.035"	140-150	18-19
75% Ar 25% CO_2 @ 25-35 CFH			
MATERIAL			
1 pc. ¼" × 6" × 10" carbon steel			
PASSES	BEAD	TIME (SEE INSTRUCTOR)	
Multiple	Weave		

Procedure 90

Running a Flat Pad of Stringer Beads

OBJECTIVE

Upon completion of this lesson you should be able to weld a pad of stringer beads in the flat position.

Text Reference: Section V, Lesson 4A; page 369.

PROCEDURE

1. Adjust the power supply and wire feeder to obtain 18 to 19 volts and 140 to 150 amperes, gas flow 25 to 35 CFH.
2. Using the forehand technique (see Figure 90-1) begin at the right side of the plate and deposit a weld bead approximately ½ in. from the edge of the plate. (See Figure 90-2.) Travel at an even speed. This will produce a bead with even width and reinforcement. Use one of the weaves shown in Figure 90-3 to spread the weld bead. This will prevent excessive reinforcement and help make sure the toe of the weld merges smoothly with the base metal.
3. Before stopping the weld be sure you fill the crater by using a circular motion of the gun over the puddle to fill the crater and reinforce it. When you extinguish the arc hold the gun over the crater to allow the post flow of gas to protect the hot weld bead from the atmosphere.
4. Continue to deposit overlapping beads on the plate, as shown in Figure 90-4. Cool the plate as necessary.
5. When finished the beads should be straight, of even width and height, and have a smooth contour.
6. Weld another plate, using the same welding conditions and bead sequence, with the backhand technique. Note the difference in bead appearance and buildup.

206 GAS METAL ARC WELDING PROCESS

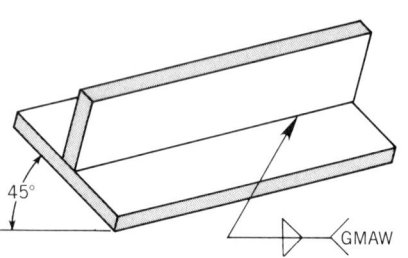

FIGURE 91-1

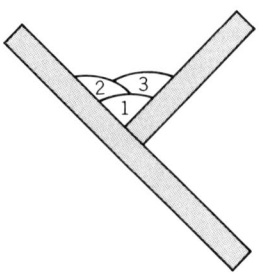

FIGURE 91-2

NAME	Fillets	POSITION	Flat
ELECTRODE E-70S-3 75% Ar 25% CO$_2$ @ 25-35 CFH	DIAMETER .035"	AMPERAGE 140-150	VOLTAGE 18-19
MATERIAL	1 pc. ¼" × 1½" × 8" carbon steel 1 pc. ¼" × 3" × 8" carbon steel		
PASSES Multiple	BEAD String	TIME (SEE INSTRUCTOR)	

Procedure 91

T-Joint Fillet, 1F Position

OBJECTIVE

Upon completion of this lesson you should be able to weld fillets in the flat position.

Text Reference: Section V, Lesson 4A; page 369.

PROCEDURE

1. Adjust the power supply and wire feeder to obtain 18 to 19 volts and 140 to 150 amperes, gas flow 25 to 35 CFH.
2. Clean the pieces to be joined. The areas to be welded should be thoroughly cleaned of any mill scale or rust. The GMAW process does not have the ability to remove many impurities because it does not have a flux. Clean weld joints will result in higher quality welds.
3. Tack the pieces together and position them as shown in Figure 91-1. Hold the gun perpendicular to the joint and using the forehand technique and one of the weaves, weld a bead from right to left. Use an even travel speed and watch for even reinforcement and good tie in of the toe of the weld to the base metal.

Remember: When stopping, fill and cool the crater properly.

4. Complete the first weld bead and apply two more beads as shown in Figure 91-2. Cool the plate between beads.
5. When you have completed the three bead fillet, cool the T assembly and inspect it. The legs should be of equal length. The toe of the weld should merge smoothly with the base material. The face of the weld should be nearly flat with no excessive concavity or convexity.
6. Weld the second side of the T assembly using the backhand technique. Note the difference in bead appearance and buildup. Use the same weave techniques and bead placement you used on the first side. When you have completed it, inspect it as you did the first side.

208 GAS METAL ARC WELDING PROCESS

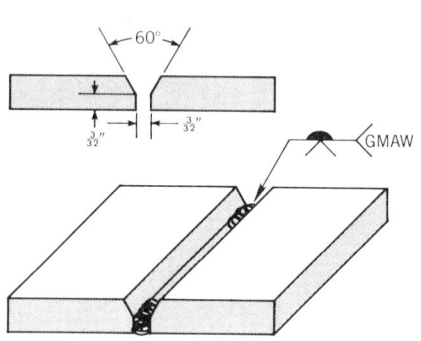

FIGURE 92-1

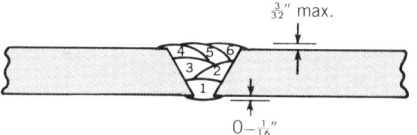

FIGURE 92-2

NAME Grooves	POSITION Flat	
ELECTRODE E-70S-3 DIAMETER .035" 75% Ar 25% CO_2 @ 25-35 CFH	AMPERAGE 125-150	VOLTAGE 18-19
MATERIAL 2 pcs. ⅜" × 3" × 6" carbon steel with a 30 degree bevel on one side		
PASSES Multiple BEAD String	TIME (SEE INSTRUCTOR)	

Procedure 92

V-Groove Butt Joint, 1G Position

OBJECTIVE

Upon completion of this lesson you should be able to weld flat grooves.

Text Reference: Section V, Lesson 4A; page 370.

PROCEDURE

1. Adjust the power supply and wire feeder to obtain 18 to 19 volts and 125 to 150 amperes, gas flow 25 to 35 CFH.
2. Thoroughly clean the pieces to be joined. Pay particular attention to the top of the plate, the side walls of the groove, and the underside of the joint. Grind or file a 3/32 in. root face on each beveled piece. (See Figure 92-1.)
3. Tack the pieces together and position as shown in Figure 92-1. Put spacers under the plate so you do not weld the plate to your table.
4. Using the backhand technique hold the gun perpendicular to the joint and strike the arc at the tack. Weave the gun from side to side. As you weave watch closely when you are in the center of the joint. By concentrating the arc on the leading edge of the puddle you can cause the bead to penetrate through the joint and fuse both root faces. You must be very careful. If you bring the arc too far down the puddle, the wire will go through the joint and the arc will become very erratic. If you allow the arc to go too far up on the puddle your penetration will decrease and you will not penetrate the joint. Practice will help you use the arc to control the flow of the weld puddle.
5. Complete the joint using the bead sequence shown in Figure 92-2. Use a slight weave to help the weld flow and to fuse to the side walls of the groove and the previous beads.
6. When you have completed the plate, cool and examine it. The root should show fill penetration along the entire length. The root face should protrude beyond the joint from 0 to 1/16 in. The face of the weld should merge smoothly with the base metal. The face should be at least flush with the base metal and not exceed 3/32 in. reinforcement.

210 GAS METAL ARC WELDING PROCESS

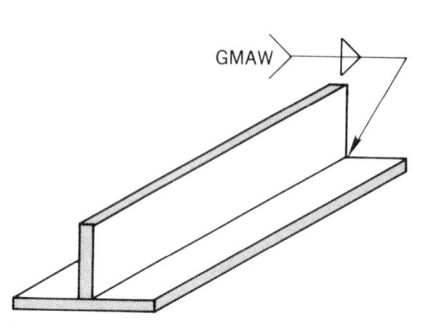

FIGURE 93-1

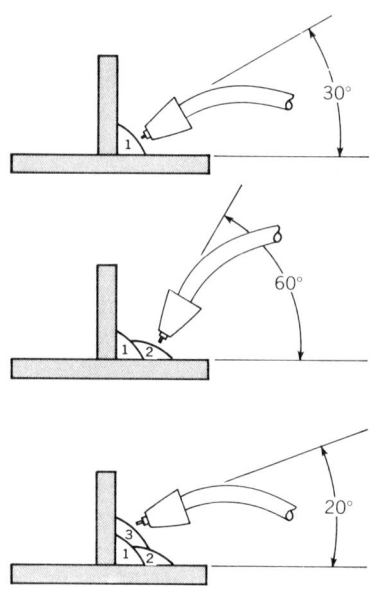

FIGURE 93-2

NAME Fillets	POSITION Horizontal	
ELECTRODE E-70S-3 DIAMETER .035" 75% Ar 25% CO_2 @ 25-35 CFH	AMPERAGE 140-150	VOLTAGE 18-19
MATERIAL 1 pc. ¼" × 1 ½" × 8" carbon steel 1 pc. ¼" × 3" × 8" carbon steel		
PASSES Multiple BEAD Weave	TIME (SEE INSTRUCTOR)	

Procedure 93

T-Joint Fillets, 2F Position

OBJECTIVE

Upon completion of this lesson you should be able to weld horizontal fillets.

Text Reference: Section V, Lesson 4B; page 371.

PROCEDURE

1. Adjust the power supply and wire feeder to obtain 18 to 19 volts, 140 to 150 amperes, gas flow 25 to 35 CFH.
2. Clean the pieces to be joined, tack them together and position them as shown in Figure 93-1.
3. Use the forehand technique, beginning at the right side of the joint and progress to the left. Use a slight weave, hesitating on the vertical part of the T joint. Use the gun angles shown in Figure 93-2.
4. Deposit two more beads using the bead sequence and gun angles shown in Figure 93-2.
5. Cool the T-bar assembly and examine the weld. The toe of the weld should merge smoothly with the base metal. The legs should be equal and the face of the weld should be nearly flat.
6. Weld the second side of the T-bar assembly using the backhand technique.

212 GAS METAL ARC WELDING PROCESS

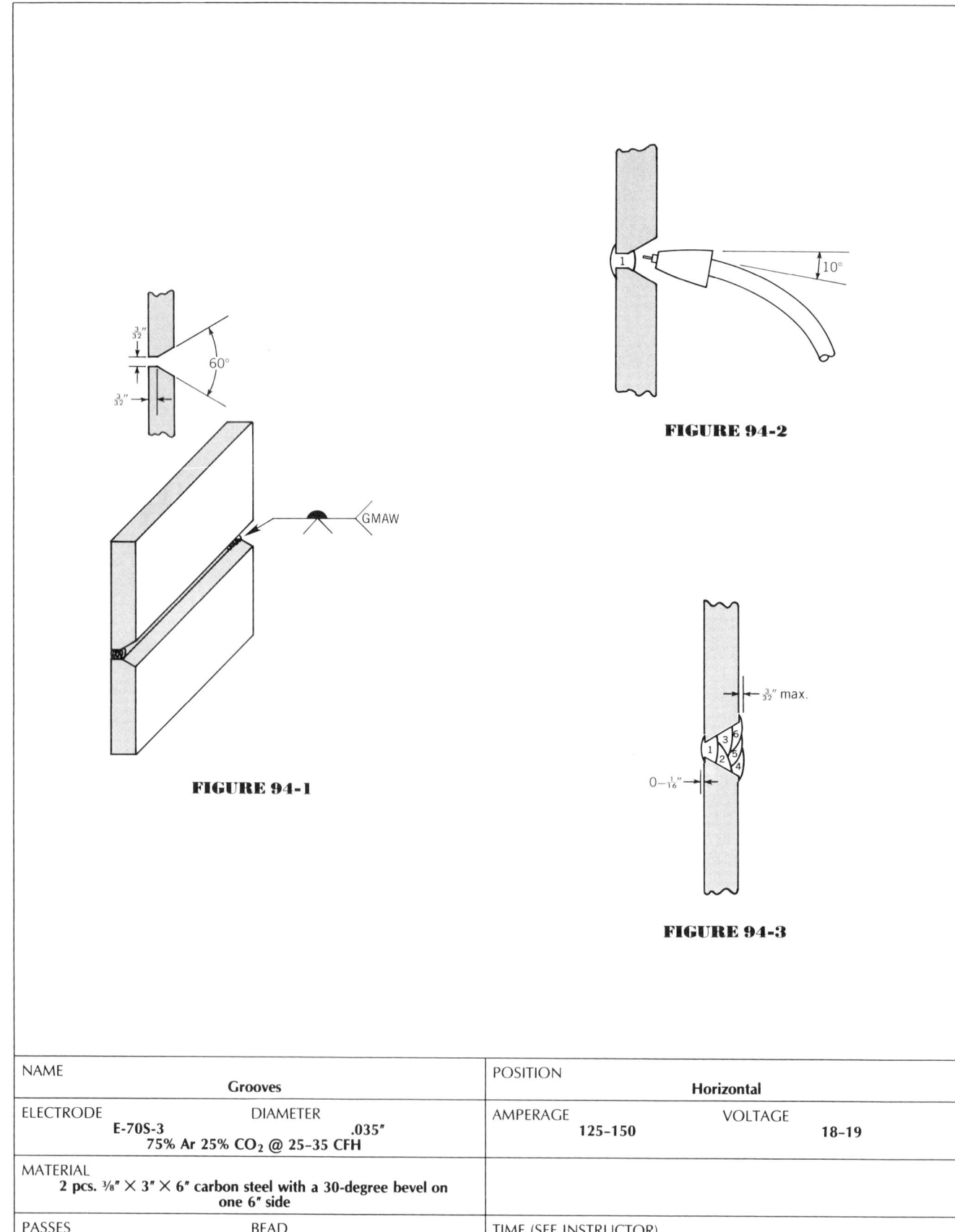

FIGURE 94-1

FIGURE 94-2

FIGURE 94-3

NAME Grooves		POSITION Horizontal	
ELECTRODE DIAMETER E-70S-3 .035" 75% Ar 25% CO$_2$ @ 25–35 CFH		AMPERAGE 125–150	VOLTAGE 18–19
MATERIAL 2 pcs. 3/8" × 3" × 6" carbon steel with a 30-degree bevel on one 6" side			
PASSES BEAD Multiple Weave		TIME (SEE INSTRUCTOR)	

Procedure 94

V-Groove Butt Joint, 2G Position

OBJECTIVE

Upon completion of this lesson you should be able to weld horizontal grooves.

Text Reference: Section V, Lesson 4B; page 372.

PROCEDURE

1. Adjust the power supply and wire feeder to obtain 18 to 19 volts and 125 to 150 amperes, gas flow 25 to 35 CFH.
2. Thoroughly clean the pieces to be joined, tack them together and position them as shown in Figure 94-1.
3. Use the backhand technique beginning at the left side of the plate. Use the arc on the puddle to push the weld through the joint to penetrate the root. Use a slight weaving motion, hesitating on the top plate. Use a gun angle as shown in Figure 94-2.
4. Complete the plate using the bead sequence shown in Figure 94-3. Cool the plate and examine it for penetration, reinforcement, and bead appearance.
5. Weld another joint using the backhand technique for the root pass (first pass) and the forehand technique for the rest of the beads.

214 GAS METAL ARC WELDING PROCESS

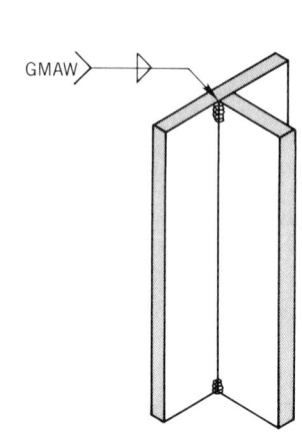

FIGURE 95-1

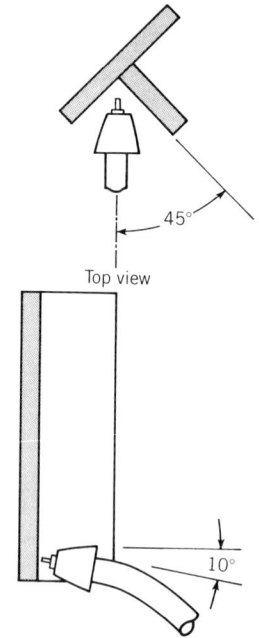

FIGURE 95-2

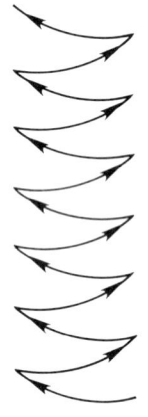

FIGURE 95-3

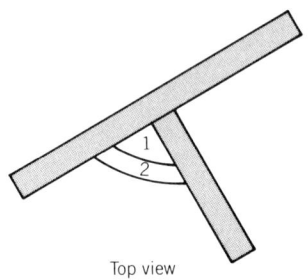

FIGURE 95-4

NAME	Fillets	POSITION	Vertical up
ELECTRODE E-70S-3 DIAMETER .035" 75% Ar 25% CO₂ @ 25-35 CFH		AMPERAGE 125-150	VOLTAGE 18-19
MATERIAL 1 pc. ¼" × 1½" × 8" carbon steel 1 pc. ¼" × 3" × 8" carbon steel			
PASSES Multiple	BEAD Weave	TIME (SEE INSTRUCTOR)	

Procedure 95

T-Joint Fillet, 3F Position, Upward Welding

OBJECTIVE

Upon completion of this lesson you should be able to weld vertical up fillets.

Text Reference: Section V, Lesson 4C; page 373.

PROCEDURE

1. Adjust the power supply and wire feeder to obtain 18 to 19 volts and 125 to 150 amperes, gas flow 25 to 35 CFH. Stay on the low side of the range for vertical welding.
2. Thoroughly clean the pieces to be joined, tack them together and position them as shown in Figure 95-1.
3. Beginning at the bottom of the joint, use the gun angles shown in Figure 95-2 and begin to weld using a weaving motion similar to that shown in Figure 95-3.
4. The weld will build a shelf at the bottom of the joint which you can build on. Make sure when you weave the torch that the arc reaches the root of the joint. This ensures good root penetration. Hesitate on the sides to fill in the weld and prevent undercut. Increase your travel speed when going from side to side to prevent excessive buildup, which would result in a very convex bead.
5. Complete the first pass, keeping the fillet size to as close to ¼ in. as possible.
6. Cool the plate thoroughly and deposit a second pass. Keep the second pass fillet size to ⅜ in. (See Figure 95-4.)
7. Weld the second side of the T-bar assembly using the same technique as the first side.

216 GAS METAL ARC WELDING PROCESS

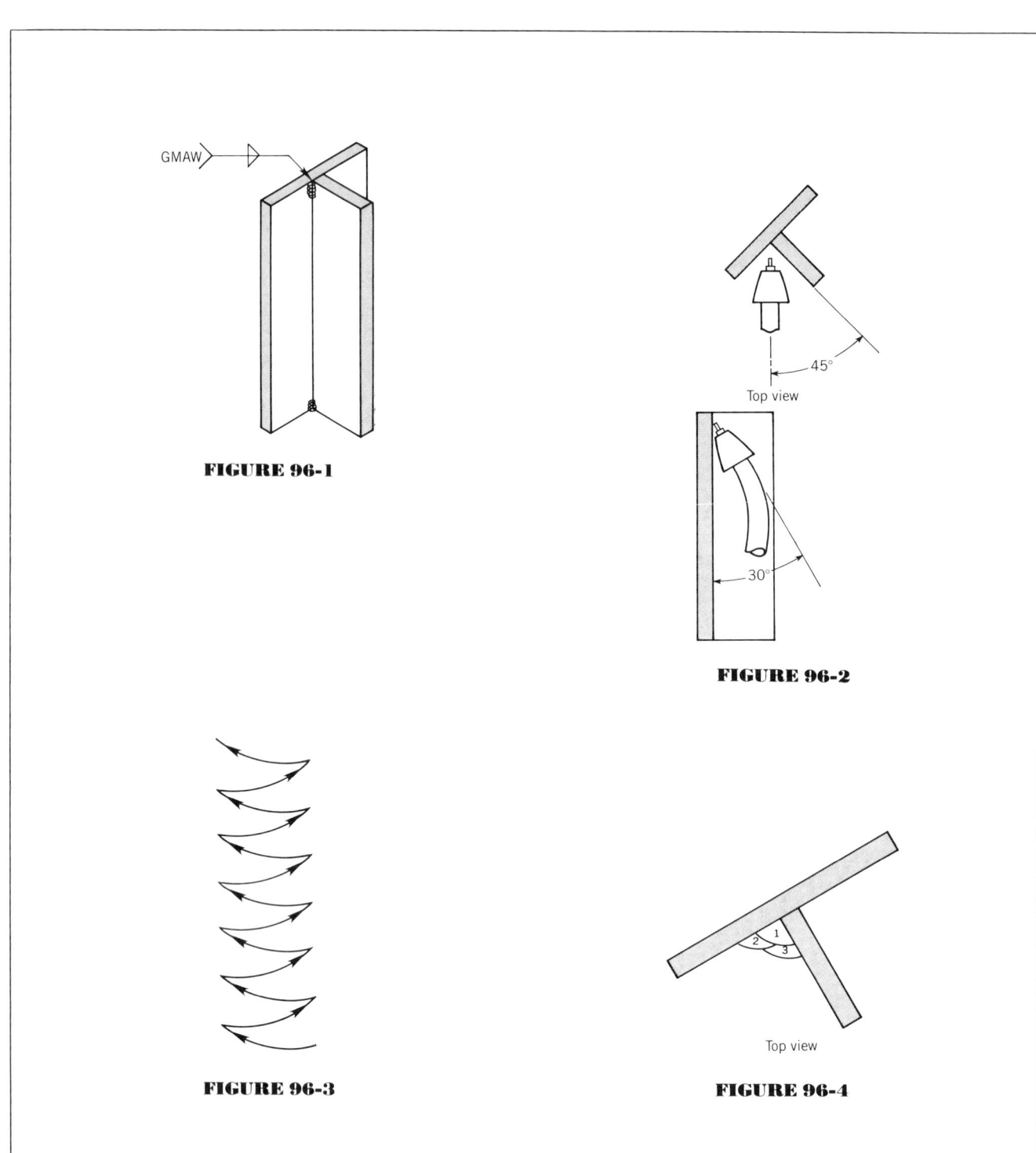

FIGURE 96-1

FIGURE 96-2

FIGURE 96-3

FIGURE 96-4

NAME	Fillets	POSITION	Vertical down
ELECTRODE E-70S-3	DIAMETER .035"	AMPERAGE 125–150	VOLTAGE 18–19
75% Ar 25% CO_2 @ 25–35 CFH			
MATERIAL	1 pc. ¼" × 1½" × 8" carbon steel 1 pc. ¼" × 3" × 8" carbon steel		
PASSES Multiple	BEAD Weave	TIME (SEE INSTRUCTOR)	

Procedure 96

T-Joint Fillet, 3F Position, Downward Welding

OBJECTIVE

Upon completion of this lesson you should be able to weld vertical down fillets.

Text Reference: Section V, Lesson 4C; page 374.

PROCEDURE

1. Adjust the power supply and wire feeder to obtain 18 to 19 volts and 125 to 150 amperes, gas flow 25 to 35 CFH.
2. Clean the pieces to be cleaned, tack them together and position them as shown in Figure 96-1.
3. Begin at the top of the joint. Use the gun angles shown in Figure 96-2. Use a slight weave pattern like the one shown in Figure 96-3. Some weave is necessary to ensure fusing along the toe of the weld. It is possible to get a condition known as cold lap* in vertical down welding. If you do not weave the gun to make sure the arc melts the side walls and fuses the weld and base metal, the arc will dwell on the puddle which will cushion the force of the arc and cause lack of fusion.
4. Continue down the joint. Your speed of travel must be fast enough to stay ahead of the weld puddle so the arc can melt the base metal. If you do not travel fast enough you will lose penetration and eventually the puddle will sag and fall out of the joint.
5. Apply two more beads, as shown in Figure 96-4. Make sure you use a slight weave to ensure side wall penetration and fusion.
6. Weld the second side using the same techniques you learned on the first side. Inspect both sides for leg length, bead appearance, and fusion at the toe of the weld.

*Cold lap is a condition where the metal lies against another surface, without melting and fusing to it.

218 GAS METAL ARC WELDING PROCESS

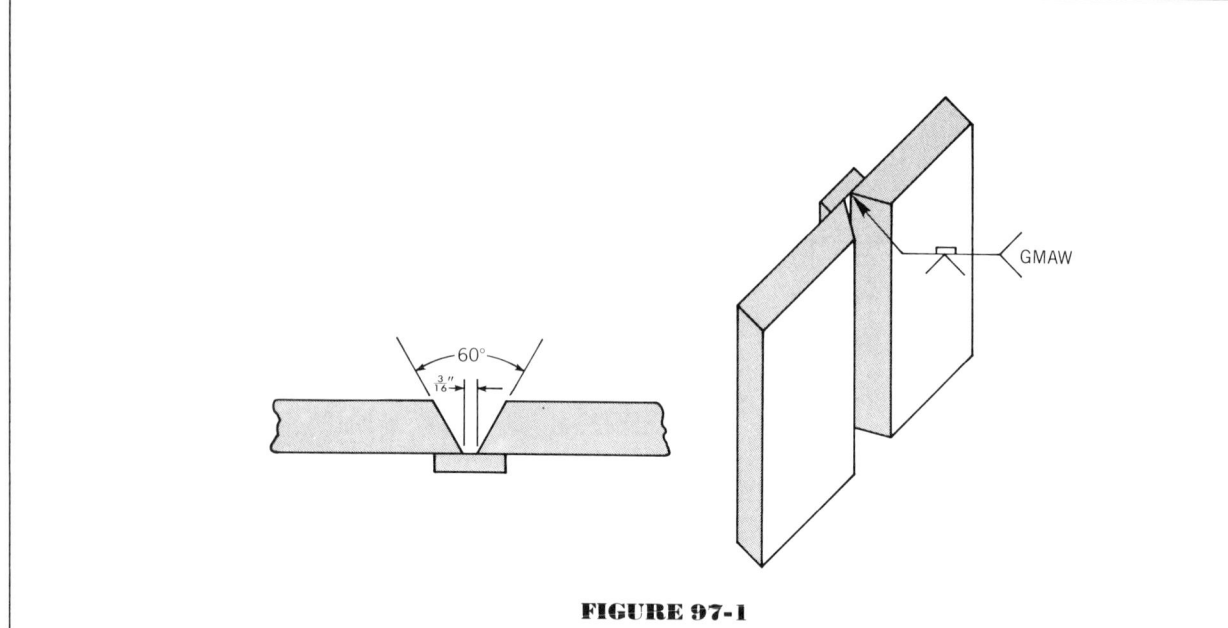

FIGURE 97-1

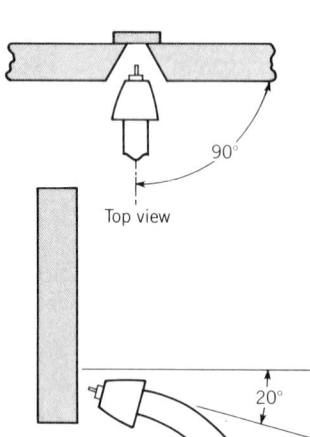

FIGURE 97-2

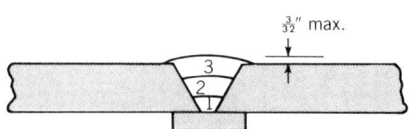

FIGURE 97-3

NAME Grooves	POSITION Vertical up
ELECTRODE E-70S-3 DIAMETER .035" 75% Ar 25% CO_2 @ 25-35 CFH	AMPERAGE 120-140 VOLTAGE 18-19
MATERIAL 2 pcs. 3/8" × 3" × 8" carbon steel with a 30-degree bevel on one 6" side 1 pc. 1/4" × 1" × 6" carbon steel	
PASSES Multiple BEAD Weave	TIME (SEE INSTRUCTOR)

Procedure 97

V-Groove Butt Joint, 3G Position with Backup Bar, Upward Welding

OBJECTIVE

Upon completion of this lesson you should be able to weld vertical up grooves.

Text Reference: Section V, Lesson 4C; page 374.

PROCEDURE

1. Adjust the power supply and wire feeder to obtain 18 to 19 volts and 120 to 140 amperes, gas flow 25 to 35 CFH. Stay on the low side of the range.
2. Clean the pieces to be joined, tack them together and position them as shown in Figure 97-1. Thoroughly tack the backing bar to the plates.
3. Begin at the bottom of the groove using the gun positions shown in Figure 97-2. Build a shelf for the weld and using a slight weave motion continue the weld up the joint. Make sure the weave is wide enough to cause the arc to melt the root of the joint ensuring good penetration and fusion.
4. After completing the root pass, deposit two more layers using a single bead weave to produce each layer. (See Figure 97-3.) Increase the width of the weave for each layer. The second layer should be just under flush. This will allow you to reinforce the joint properly. Hesitate on each side when weaving to fill in the weld and prevent undercut. Cool the plate between each layer.

220 GAS METAL ARC WELDING PROCESS

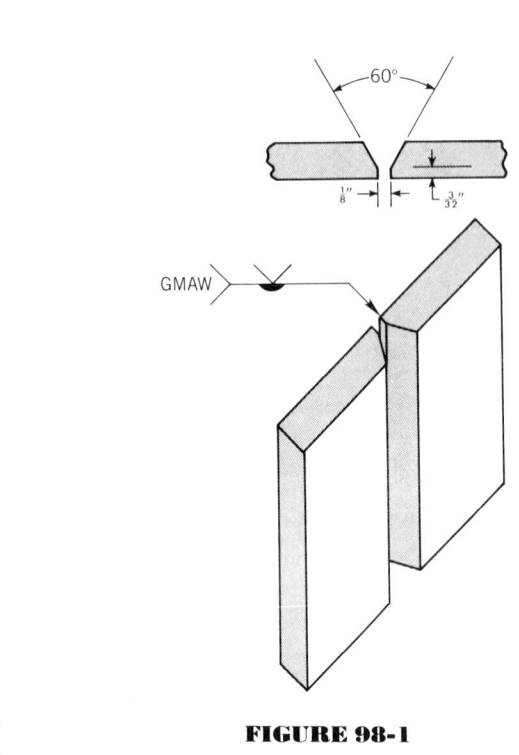

FIGURE 98-1

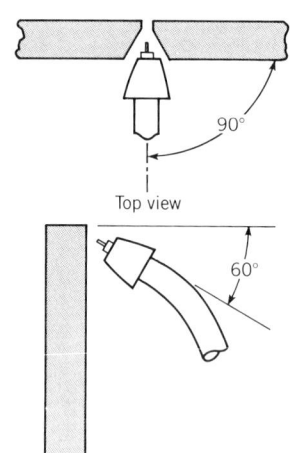

FIGURE 98-2

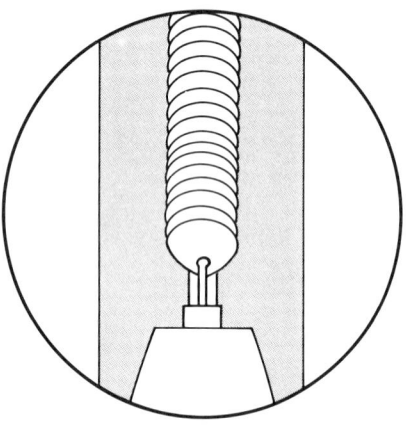

FIGURE 98-3

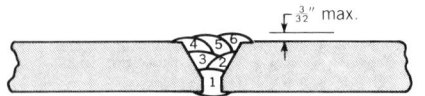

FIGURE 98-4

NAME		POSITION	
	Grooves		Vertical down
ELECTRODE	DIAMETER	AMPERAGE	VOLTAGE
E-70S-3	.035"	120–140	18–19
75% Ar 25% CO₂ @ 25–35 CFH			
MATERIAL			
2 pcs. ³⁄₈" × 3" × 6" carbon steel with a 30-degree bevel on one 6" side			
PASSES	BEAD	TIME (SEE INSTRUCTOR)	
Multiple	Weave		

Procedure 98

V-Groove Butt Joint, 3G Position, Downward Welding

OBJECTIVE

Upon completion of this lesson you should be able to weld vertical down grooves.

Text Reference: Section V, Lesson 4C; page 375.

PROCEDURE

1. Adjust the power supply and wire feeder to obtain 18 to 19 volts and 120 to 140 amperes, gas flow 25 to 35 CFH.
2. Clean the pieces to be welded. Grind or file a 3/32 in. root face on each plate bevel. Tack and position the pieces as shown in Figure 98-1.
3. Start at the top using the gun positions shown in Figure 98-2. Use a side-to-side weaving technique. A weld will form across the root opening. Carefully manipulate the gun side to side to get side wall fusion. Keep the tip of the wire on the very leading edge of the puddle and deep into the joint. (See Figure 98-3.) This will cause the root of the weld to penetrate beyond the root of the joint and give a convex root reinforcement. Gauge your travel speed to keep ahead of the puddle.
4. Continue with the root pass to the end of the joint. Cool the plate and add more passes to complete the joint using the sequence shown in Figure 98-4. Use a slight weave on each bead.
5. Note: It takes more passes to fill the joint. To prevent weld puddle sag it is necessary to keep the wire on the leading edge of the puddle. This results in fast travel speeds which keeps deposition rates low. Whenever welding vertical down, it is very important to keep the wire on the lead edge of the puddle and to weave the puddle enough to point the arc to those points you want to weld. Vertical down, short-circuiting GMAW does not have a "hot" arc. Gun manipulation is extremely important to prevent lack of fusion defects.

222 GAS METAL ARC WELDING PROCESS

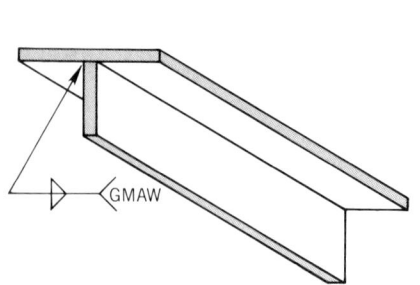

FIGURE 99-1

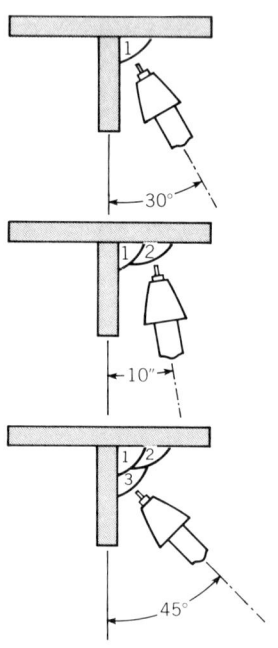

FIGURE 99-2

NAME	Fillets	POSITION	Overhead
ELECTRODE E-70S-3	DIAMETER .035″ 75% Ar 25% CO_2 @ 25–35 CFH	AMPERAGE 125–150	VOLTAGE 18–19
MATERIAL	1 pc. ¼″ × 1 ½″ × 8″ carbon steel 1 pc. ¼″ × 3″ × 8″ carbon steel		
PASSES Multiple	BEAD Weave	TIME (SEE INSTRUCTOR)	

Procedure 99

T-Joint Fillet, 4F Position

OBJECTIVE

Upon completion of this lesson you should be able to weld overhead fillets.

Text Reference: Section V, Lesson 4D, page 376.

PROCEDURE

1. Adjust the power supply and wire feeder to obtain 18 to 19 volts and 125 to 150 amperes, gas flow 25 to 35 CFH.
2. Clean the pieces to be welded, tack them together, and position them as shown in Figure 99-1.
3. Use the gun angles shown in Figure 99-2 and the forehand technique to put in the first pass. Slightly weave the gun to prevent lack of fusion along the toe of the weld.
4. Cool the plate, clean it, and put in two more passes using the gun angles shown in 99-2.
5. Cool the plate and weld the second side of the T assembly using the backhand technique and the gun angles shown in Figure 99-2.

224 GAS METAL ARC WELDING PROCESS

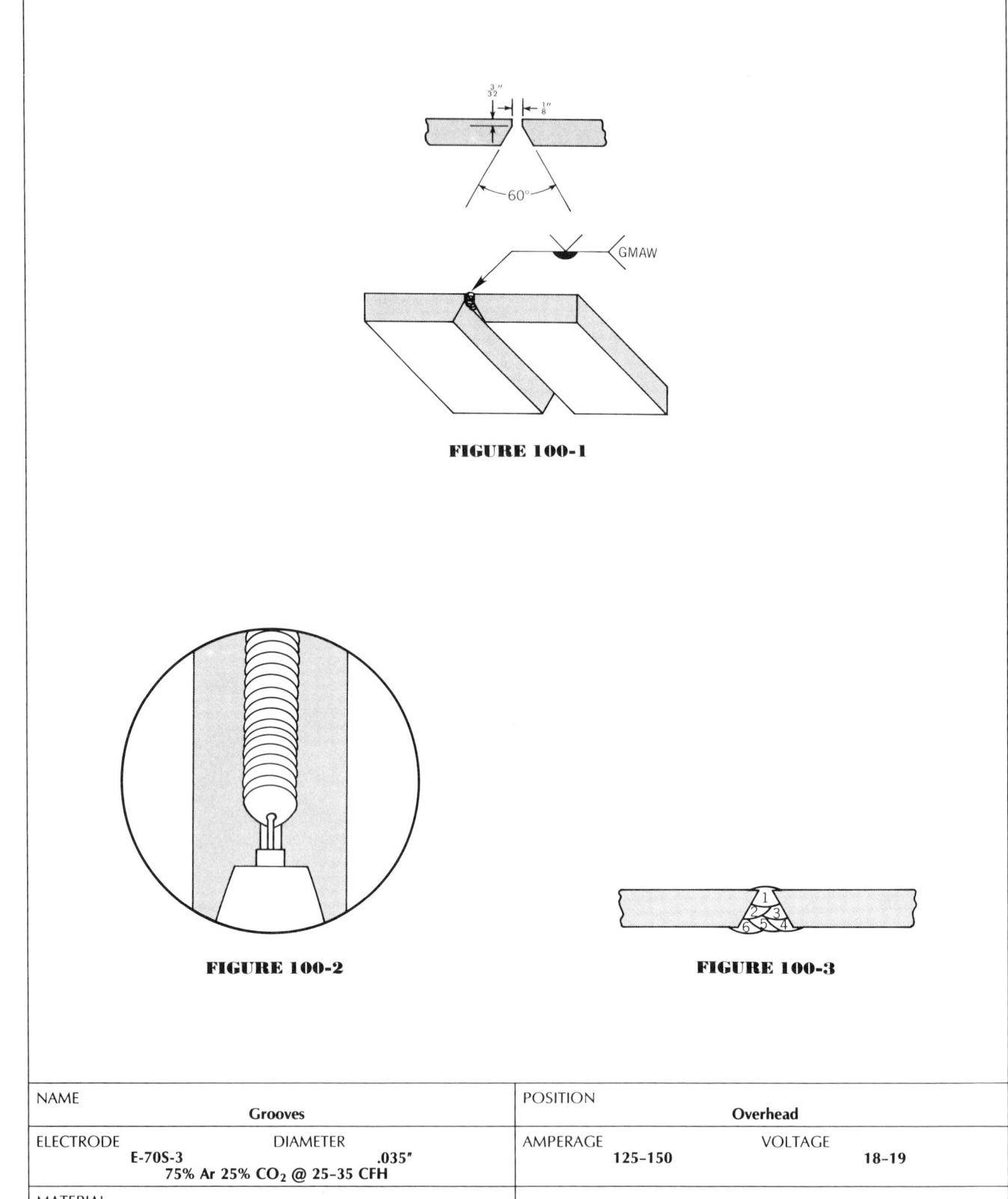

FIGURE 100-1

FIGURE 100-2

FIGURE 100-3

NAME Grooves	POSITION Overhead
ELECTRODE E-70S-3 DIAMETER .035" 75% Ar 25% CO_2 @ 25-35 CFH	AMPERAGE 125–150 VOLTAGE 18–19
MATERIAL 2 pcs. ⅜" × 3" × 6" carbon steel with a 30-degree bevel on one side	
PASSES Multiple BEAD Weave	TIME (SEE INSTRUCTOR)

Procedure 100

V-Groove Butt Joint, 4G Position

OBJECTIVE

Upon completion of this lesson you should be able to weld overhead grooves.

Text Reference: Section V, Lesson 4D; page 376.

PROCEDURE

1. Adjust the power supply and wire feeder to obtain 18 to 19 volts and 125 to 150 amperes, gas flow 25 to 35 CFH.
2. Clean the pieces to be welded. File or grind a 3/32 in. root face on the bevel. Tack the pieces together and position them as shown in Figure 100-1.
3. Keep the gun perpendicular to the joint using a slight backhand technique and a weave. Begin at the end of the joint and keep the wire at the leading edge of the puddle. The position of the wire will determine the amount of penetration and shape of the root pass. Keep the wire on the leading edge of the puddle and deep into the joint as you did in the previous lesson. (See Figure 100-2.) This will push the weld puddle through the joint to produce convex root face with reinforcement beyond the root of the joint.
6. Complete the joint using the bead sequence shown in Figure 100-3. Change the gun angle for each bead to point in the direction of the bead. Weave each bead to ensure good penetration with the side walls of the groove, and other weld beads.

Spray Transfer Welding of Steel

In this section you will practice welding carbon steel with the gas metal arc process using the spray transfer. As you have learned there are several gases or gas mixes that can be used for shielding. For the lessons in this section we will be using 98% argon 2% oxygen. The arc in GMAW spray transfer is considerably hotter, producing more infrared and ultraviolet radiation than the short-circuiting GMAW process. Precaution must be taken to protect yourself and others in the area from this high intensity arc. Painful burns can result from short exposure to these rays.

GENERAL INSTRUCTIONS

Set up your equipment as you did for short arc welding, however, recess your tip 1/8 in. below the lip of the cup. If you do not you will stand a good chance of burning the wire back to the tip and melting it off. The wire size we will use for the procedures in this section is .045 in.

MATERIALS AND EQUIPMENT

For this section the material and equipment will be as follows:

1. Welding shield with appropriate lens
2. Safety glasses
3. Gauntlet type leather gloves
4. Appropriate welding clothing
5. Wire brush
6. Diagonal side cutters

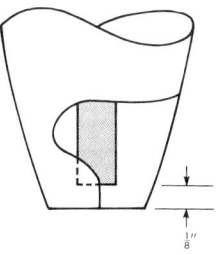

228 GAS METAL ARC WELDING PROCESS

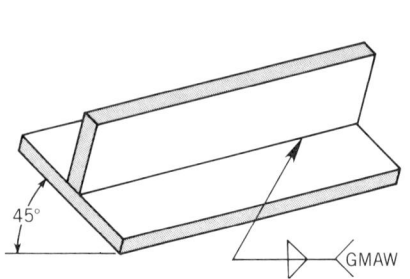

FIGURE 101-1

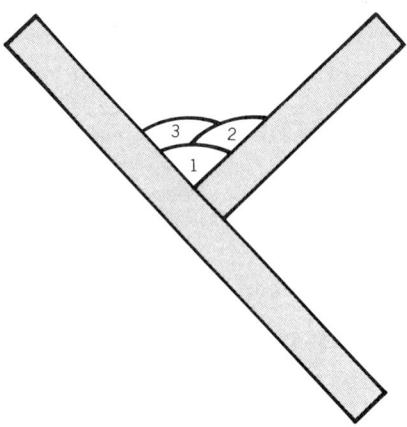

FIGURE 101-2

NAME Fillet	POSITION Flat	
ELECTRODE E-70S-3 100% Ar @ 25-35 CFH	DIAMETER .035"	AMPERAGE 190–210	VOLTAGE 24–25
MATERIAL 1 pc. ¼" × 1 ½" × 8" carbon steel 1 pc. ¼" × 3" × 8" carbon steel		
PASSES Multiple	BEAD String	TIME (SEE INSTRUCTOR)

Procedure 101

T-Joint Fillet, 1F Position

OBJECTIVE

Upon completion of this lesson you should be able to weld flat fillets.

Text Reference: Section V, Lesson 5A; page 379.

PROCEDURE

1. Adjust the power supply and wire feeder to obtain 24 to 25 volts and 190 to 210 amperes, gas flow 25 to 35 CFH.
2. Clean the pieces to be welded, tack them together, and position them as shown in Figure 101-1.
3. Using the forehand technique begin at the right side of the T assembly and weld toward the left. Note that the wire does not contact the plate and the puddle is very fluid and flows well thus making weaving unnecessary.
4. Watch the root of the joint to make sure the arc melts the metal completely. Regulate your travel accordingly. Move the gun only when melting is complete. Keep the gun perpendicular to the joint.
5. Apply a three-pass weld using the bead sequence shown in Figure 101-2. When you are through with the first side cool the plate and weld the second side using the backhand technique. Use the same bead sequence you did on the first side.

230 GAS METAL ARC WELDING PROCESS

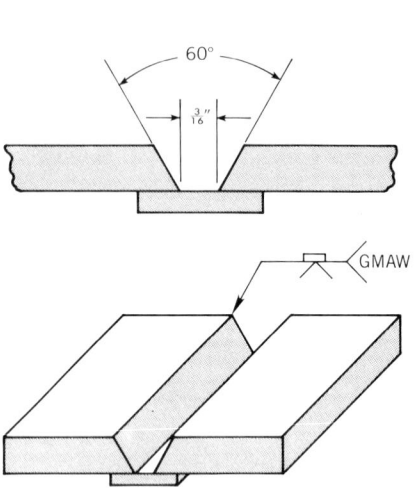

FIGURE 102-1

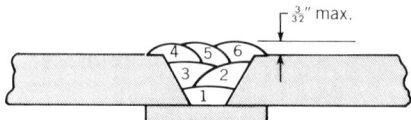

FIGURE 102-2

NAME	Groove	POSITION	Flat
ELECTRODE E-70S-3	DIAMETER .035" 100% Ar @ 25-35 CFH	AMPERAGE 190-210	VOLTAGE 24-25
MATERIAL	2 pcs. 3/8" × 3" × 6" carbon steel 1 pc. 1/4" × 1" × 6" carbon steel		
PASSES Multiple	BEAD String	TIME (SEE INSTRUCTOR)	

Procedure 102

V-Groove Butt Joint, 1G Position with Backup Bar

OBJECTIVE

Upon completion of this lesson you should be able to weld flat grooves.

Text Reference: Section V, Lesson 5A; page 379.

PROCEDURE

1. Adjust the power supply and wire feeder to obtain 24 to 25 volts and 190 to 210 amperes, gas flow 25 to 35 CFH.
2. Clean the pieces to be welded, tack them together, and position them as shown in Figure 102-1.
3. Use the forehand technique to weld the root pass, welding from right to left. Weld the root in one pass. Keep the gun perpendicular to the joint.
4. Complete the joint using the bead sequence shown in Figure 102-2. When reinforcing the weld pay attention to the toe of the weld. It is possible, when using spray transfer, to undercut the base metal at the toe of the weld. To prevent this, either slow down your travel speed or use a slight weave.
5. Prepare and weld another joint using the backhand technique. Use the bead sequence shown in Figure 102-2.

232 GAS METAL ARC WELDING PROCESS

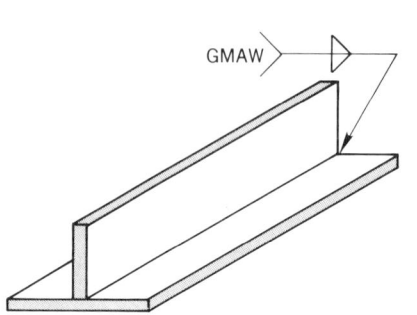

FIGURE 103-1

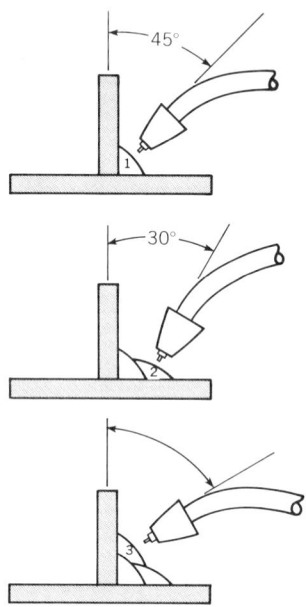

FIGURE 103-2

NAME	Fillet	POSITION	Horizontal	
ELECTRODE E-70S-3	DIAMETER .035" 100% Ar @ 25-35 CFH	AMPERAGE 190-210	VOLTAGE	24-25
MATERIAL	1 pc. ¼" × 1½" × 8" carbon steel 1 pc. ¼" × 3" × 8" carbon steel			
PASSES Multiple	BEAD String	TIME (SEE INSTRUCTOR)		

Procedure 103

T-Joint Fillet, 2F Position

OBJECTIVE

Upon completion of this lesson you should be able to weld horizontal fillets.

Text Reference: Section V, Lesson 5B; page 380.

PROCEDURE

1. Adjust the power supply and wire feeder to obtain 24 to 25 volts and 190 to 210 amperes, gas flow 25 to 35 CFH.
2. Clean the pieces to be welded, tack them together, and position them as shown in Figure 103-1.
3. Using the torch angles shown in Figure 103-2, weld a three-pass fillet weld using the forehand technique. Pay close attention to the third bead. Since the weld puddle is large and very fluid you must tilt the gun to deposit the metal on the vertical part of the T assembly. Watch the toe of the weld closely and use the gun angle and travel speed to prevent undercut in this area.
4. Cool the T assembly and weld the second side of the joint using the backhand technique.

234 GAS METAL ARC WELDING PROCESS

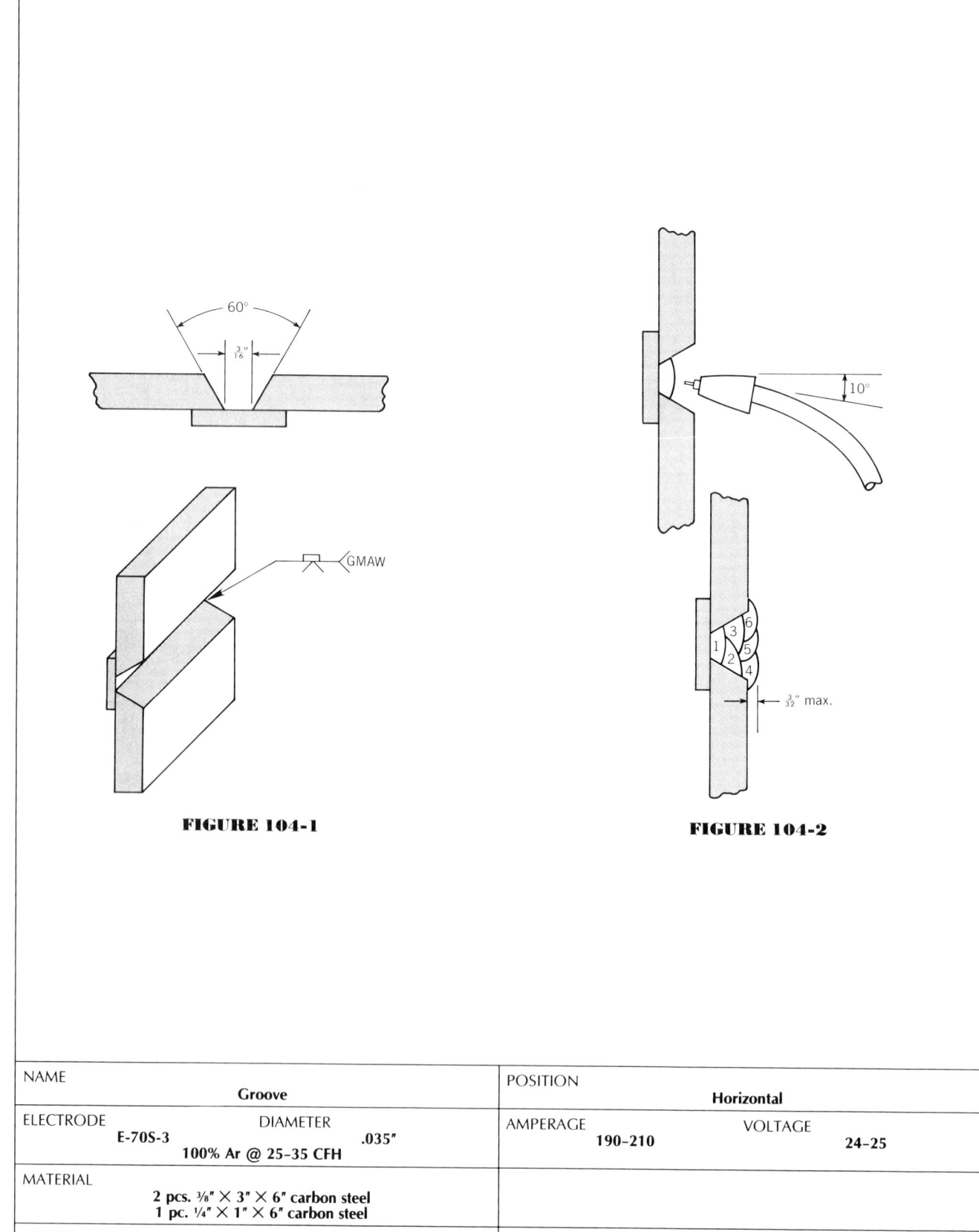

FIGURE 104-1

FIGURE 104-2

NAME	Groove	POSITION	Horizontal	
ELECTRODE E-70S-3	DIAMETER .035" 100% Ar @ 25-35 CFH	AMPERAGE 190-210	VOLTAGE	24-25
MATERIAL	2 pcs. ⅜" × 3" × 6" carbon steel 1 pc. ¼" × 1" × 6" carbon steel			
PASSES Multiple	BEAD String	TIME (SEE INSTRUCTOR)		

Procedure 104

V-Groove Butt Joint, 2G Position with Backup Bar

OBJECTIVE

Upon completion of this lesson you should be able to weld horizontal fillets.

Text Reference: Section V, Lesson 5B; page 380.

PROCEDURE

1. Adjust the power supply and wire feeder to obtain 24 to 25 volts and 190 to 210 amperes, gas flow 25 to 35 CFH.
2. Clean the pieces to be welded, tack them together, and position them as shown in Figure 104-1.
3. Use the gun angle shown in Figure 104-2 and the forehand technique to weld the root pass in one bead.
4. Alter the gun angle from 0 to 10 degrees to deposit the remaining beads in the joint. The last bead is where you must pay particular attention to prevent undercut. Use a gun angle of 10 degrees below horizontal, and watch your travel speed. Keep the arc length short.
5. Weld another assembly using the backhand technique. Use the same gun angles and bead sequence.

FLUX-CORED ARC WELDING PROCESS

Flux-Cored Arc Welding of Steel

In this section you will be welding carbon steel with the Flux-Cored Arc Welding (FCAW) process.

GENERAL INSTRUCTIONS

Before you start to weld you must check out and set up your equipment.

1. Inspect the equipment. If there is any equipment that needs repair notify your instructor.
2. Go to the power supply, and check the cable connections. Check the correct voltage for the wire you are using and set the power supply accordingly.
3. Check the work cable and turn on the power supply.
4. Proceed to the wire feeder and check for the proper size filler wire and wire feed rolls. For the lessons in this chapter you will be told to use .045 in. diameter E-71T-1 filler wire. If the wire is not threaded through the gun and cable assembly use the wire feeder switch or gun trigger to feed wire through the conduit assembly and through the gun.

Caution: If you use the gun trigger to feed the wire make sure the gun is not touching or near objects on which an arc might be struck. Remove the tip when the wire is being fed through the gun. When the wire is fed through release the wire feed button and install the contact tip. For Flux-Cored Arc Welding the tip should be even with the gas nozzle.

5. Set the gas flow rate to 30 to 40 CFH of carbon dioxide. Apply antispatter compound to the gas cup and contact tip to prevent spatter buildup.

MATERIALS AND EQUIPMENT

For this section the material and equipment will be as follows:

1. Welding shield with appropriate lens
2. Safety glasses
3. Gauntlet type leather gloves
4. Appropriate welding clothing

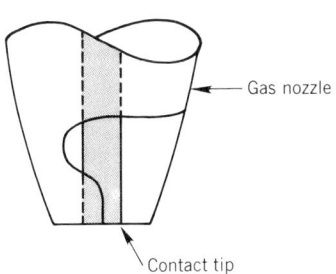

240 FLUX-CORED ARC WELDING PROCESS

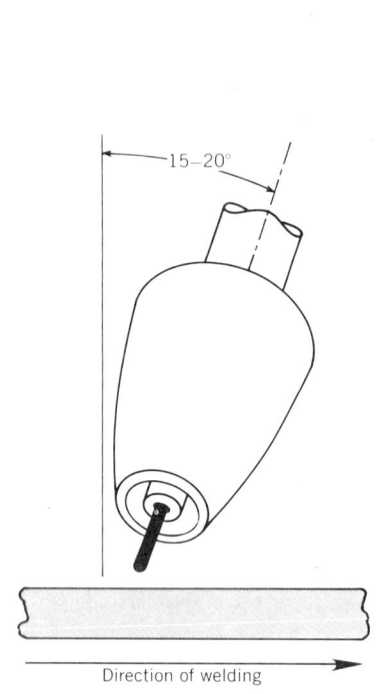

FIGURE 105-1

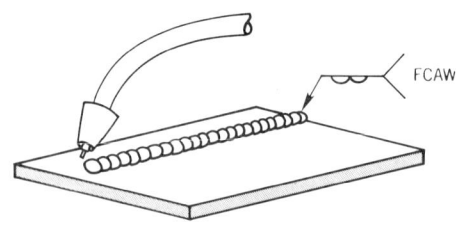

FIGURE 105-2

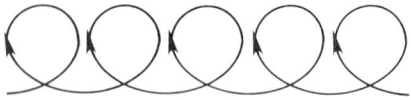

FIGURE 105-3

FIGURE 105-4

NAME	Pad of stringer beads	POSITION	Flat
ELECTRODE E-71T-1	DIAMETER .045" Gas flow 30 to 40 CFH carbon dioxide	AMPERAGE 170-190	VOLTAGE 21-23
MATERIAL	1 pc. ¼" × 6" × 10" carbon steel		
PASSES Multiple	BEAD Weave	TIME (SEE INSTRUCTOR)	

Procedure 105

Running Flat Pad of Stringers

OBJECTIVE

Upon completion of this lesson you should be able to weld a flat pad of stringer beads.

Text Reference: Section VI, Lesson 4A; page 419.

PROCEDURE

1. Adjust the power supply and wire feeder to obtain 21 to 23 volts and 170 to 190 amperes, gas flow 30 to 40 CFH.
2. Using the backhand technique (see Figure 105-1) begin at the right side of the plate and deposit a weld bead approximately ½ in. from the edge of the plate. (See Figure 105-2.) Travel at an even speed. This will produce a bead with even width and reinforcement. Use one of the weaves shown in Figure 105-3 to spread the weld bead. This will prevent excessive reinforcement and help make sure the toe of the weld merges smoothly with the base metal.
3. Before stopping the weld be sure you fill the crater by using a circular motion of the gun over the puddle to fill the crater and reinforce it. Allow the bead to cool and clean the slag from the weld.
4. Continue to deposit overlapping beads on the plate, as shown in Figure 105-4. Cool the plate as necessary.
5. When finished the beads should be straight, of even width and height, and have a smooth contour.

242 FLUX-CORED ARC WELDING PROCESS

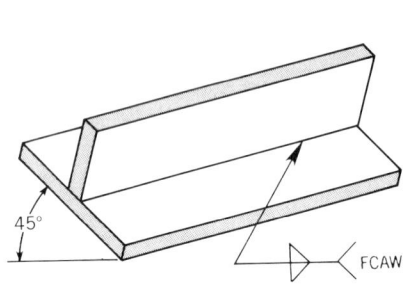

FIGURE 106-1

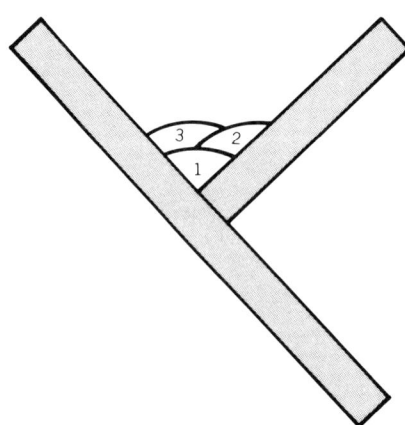

FIGURE 106-2

NAME Fillets		POSITION Flat	
ELECTRODE E-71T-1 Gas flow 30 to 40 CFH carbon dioxide	DIAMETER .045"	AMPERAGE 170-190	VOLTAGE 21-23
MATERIAL 1 pc. ¼" × 1 ½" × 8" carbon steel 1 pc. ¼" × 3" × 8" carbon steel			
PASSES Multiple	BEAD String	TIME (SEE INSTRUCTOR)	

Procedure 106

T-Joint Fillet, 1F Position

OBJECTIVE

Upon completion of this lesson you should be able to weld flat fillets.

Text Reference: Section VI, Lesson 4A; page 419.

PROCEDURE

1. Adjust the power supply and wire feeder to obtain 21 to 23 volts and 170 to 190 amperes, gas flow 30 to 40 CFH.
2. Clean the pieces to be joined. The areas to be welded should be thoroughly cleaned of any mill scale or rust. Clean weld joints will result in higher quality welds.
3. Tack the pieces together and position them as shown in Figure 106-1. Hold the gun perpendicular to the joint and using the backhand technique and one of the weaves, weld a bead from left to right. Use an even travel speed and watch for even reinforcement and good tie at the toe of the weld. Remember when stopping to fill the crater properly.
4. Complete the first weld bead, clean it and apply two more beads as shown in Figure 106-2. Cool the plate between beads.
5. When you have completed the three-bead fillet, cool the T assembly and inspect it. The legs should be of equal length. The toe of the weld should merge smoothly with the base material. The face of the weld should be nearly flat with no excessive concavity or convexity.
6. Weld the second side of the T assembly using the same technique and bead placement you used on the first side.

244 FLUX-CORED ARC WELDING PROCESS

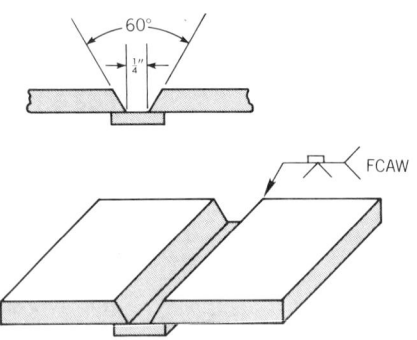

FIGURE 107-1

FIGURE 107-2

NAME	Grooves	POSITION	Flat	
ELECTRODE E-71T-1	DIAMETER .045" Gas flow 30 to 40 CFH carbon dioxide	AMPERAGE 170–190	VOLTAGE	21–23
MATERIAL	1 pc. ¼" × 1" × 6" carbon steel 2 pcs. ⅜" × 3" × 6" carbon steel with a 30-degree bevel on one 6" side			
PASSES Multiple	BEAD Weave	TIME (SEE INSTRUCTOR)		

Procedure 107

V-Groove Butt Joint, 1G Position with Backup Bar

OBJECTIVE

Upon completion of this lesson you should be able to weld flat grooves.

Text Reference: Section VI, Lesson 4A; page 420.

PROCEDURE

1. Adjust the power supply and wire feeder to obtain 21 to 23 volts and 170 to 190 amperes, gas flow 30 to 40 CFH.
2. Thoroughly clean the pieces to be joined. Pay particular attention to the top of the plate, the side walls of the groove and the underside of the joint. (See Figure 107-1.)
3. Tack the pieces together and position as shown in Figure 107-1. Tack the backing strip firmly to the plates.
4. Using the backhand technique hold the gun perpendicular to the joint and strike the arc at the tack. Weave the gun from side to side. As you weave watch closely when you are in the center of the joint. By concentrating the arc on the leading edge of the puddle you can cause the bead to penetrate into the backing strip and fuse both root faces. You must be very careful. If you allow the arc to go too far up on the puddle your penetration will decrease and you will not penetrate the joint. Practice will help you use the arc to control the flow of the weld puddle.
5. Complete the joint using the bead sequence shown in Figure 107-2. Use a slight weave to help the weld flow and to fuse to the side walls of the groove and the previous beads.
6. When you have completed the plate, cool it and examine it. The root should show fill penetration along the entire length. The root face should protrude beyond the joint from 0 to $1/16$ in. The face of the weld should merge smoothly with the base metal. The face should be at least flush with the base metal and not exceed $3/32$ in. reinforcement.

246 FLUX-CORED ARC WELDING PROCESS

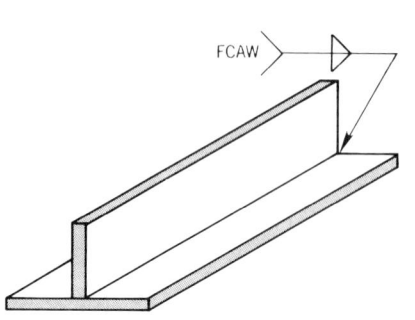

FIGURE 108-1

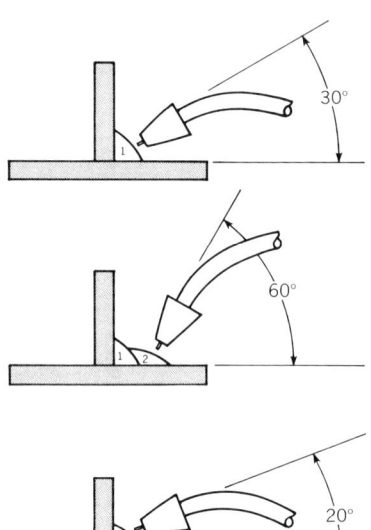

FIGURE 108-2

NAME	Fillets		POSITION	Horizontal	
ELECTRODE	E-71T-1 Gas flow 30 to 40 CFH carbon dioxide	DIAMETER .045"	AMPERAGE 170-190	VOLTAGE	21-23
MATERIAL	1 pc. ¼" × 1 ½" × 8" carbon steel 1 pc. ¼" × 3" × 8" carbon steel				
PASSES Multiple	BEAD	String	TIME (SEE INSTRUCTOR)		

Procedure 108

T-Joint Fillet, 2F Position

OBJECTIVE

Upon completion of this lesson you should be able to weld horizontal fillets.

Text Reference: Section VI, Lesson 4B; page 421.

PROCEDURE

1. Adjust the power supply and wire feeder to obtain 21 to 23 volts and 170 to 190 amperes, gas flow 30 to 40 CFH.
2. Clean the pieces to be joined, tack them together, and position them as shown in Figure 108-1.
3. Use the backhand technique, beginning at the left side of the joint and progress to the right. Use a slight weave, hesitating on the vertical part of the T joint. Use the gun angles shown in Figure 108-2.
4. Deposit two more beads using the bead sequence and gun angles shown in Figure 108-2.
5. Cool the T-bar assembly and examine the weld. The toe of the weld merge smoothly with the base metal. The legs should be equal and the face of the weld should be nearly flat.
6. Weld the second side of the T-bar assembly using the backhand technique.

248 FLUX-CORED ARC WELDING PROCESS

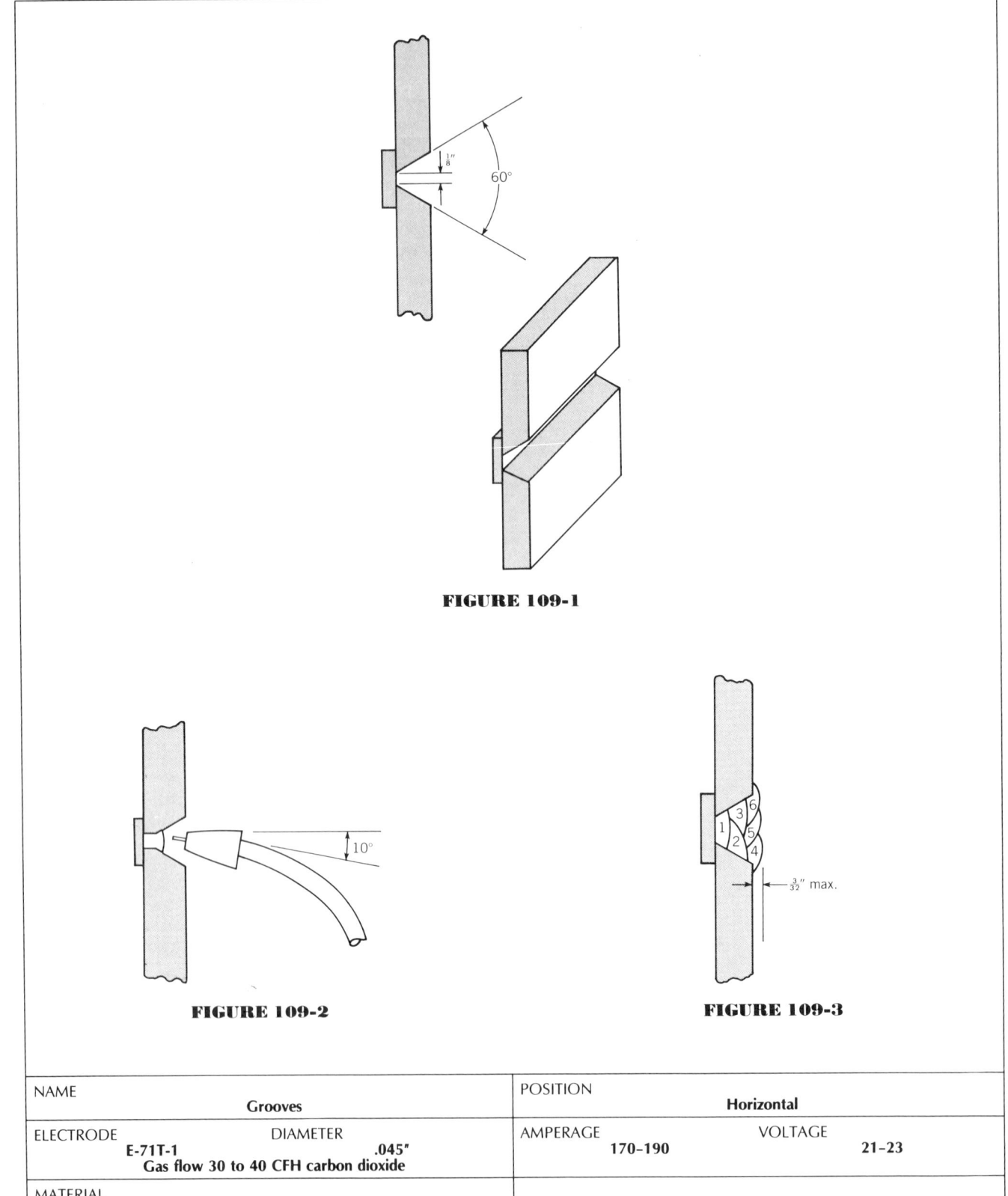

FIGURE 109-1

FIGURE 109-2

FIGURE 109-3

NAME Grooves	POSITION Horizontal	
ELECTRODE E-71T-1 Gas flow 30 to 40 CFH carbon dioxide DIAMETER .045"	AMPERAGE 170-190	VOLTAGE 21-23
MATERIAL 1 pc. ¼" × 1" × 6" carbon steel 2 pcs. ⅜" × 3" × 6" carbon steel with a 30-degree bevel on one 6" side		
PASSES Multiple BEAD String	TIME (SEE INSTRUCTOR)	

Procedure 109

V-Groove Butt Joint, 2G Position with Backup Bar

OBJECTIVE

Upon completion of this lesson you should be able to weld horizontal grooves.

Text Reference: Section VI, Lesson 4B; page 422.

PROCEDURE

1. Adjust the power supply and wire feeder to obtain 21 to 23 volts and 170 to 190 amperes, gas flow 30 to 40 CFH.
2. Thoroughly clean the pieces to be joined, tack them together, and position them as shown in Figure 109-1.
3. Use the backhand technique beginning at the left side of the plate. Use a slight weaving motion, hesitating on the top plate. Use a gun angle as shown in Figure 109-2.
4. Complete the plate using the bead sequence shown in Figure 109-3. Cool the plate and examine it for penetration, reinforcement, and bead appearance.

250 FLUX-CORED ARC WELDING PROCESS

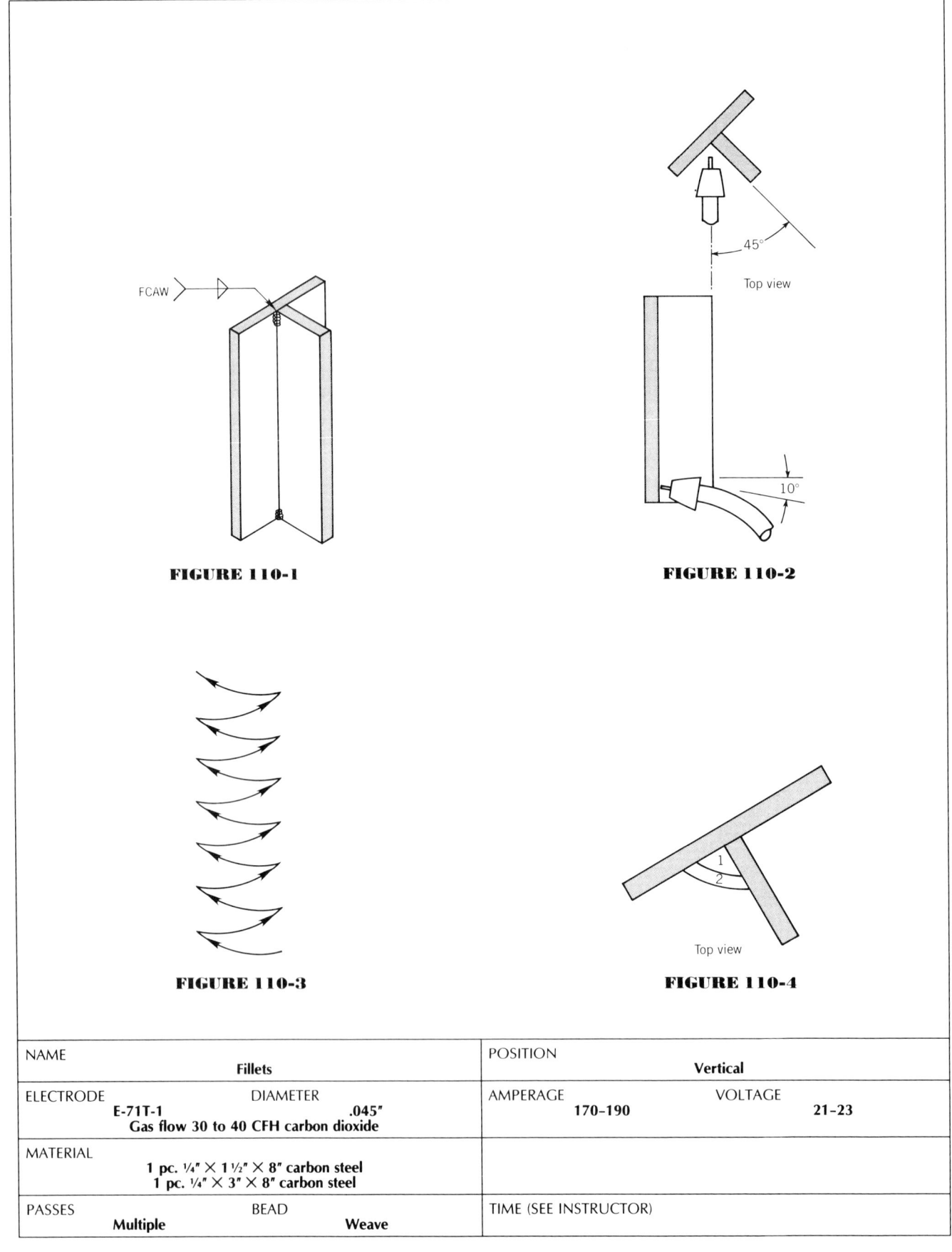

FIGURE 110-1

FIGURE 110-2

FIGURE 110-3

FIGURE 110-4

NAME	Fillets	POSITION	Vertical	
ELECTRODE E-71T-1	DIAMETER .045" Gas flow 30 to 40 CFH carbon dioxide	AMPERAGE 170-190	VOLTAGE	21-23
MATERIAL	1 pc. ¼" × 1 ½" × 8" carbon steel 1 pc. ¼" × 3" × 8" carbon steel			
PASSES Multiple	BEAD Weave	TIME (SEE INSTRUCTOR)		

Procedure 110

T-Joint Fillet, 3F Position

OBJECTIVE

Upon completion of this lesson you should be able to weld vertical up fillets.

Text Reference: Section VI, Lesson 4C; page 422.

PROCEDURE

1. Adjust the power supply and wire feeder to obtain 21 to 23 volts and 170 to 190 amperes, gas flow 30 to 40 CFH. Stay on the low side of the range for vertical welding.
2. Thoroughly clean the pieces to be joined, tack them together, and position them as shown in Figure 110-1.
3. Beginning at the bottom of the joint, use the gun angles shown in 110-2 and begin to weld using a weaving motion similar to that shown in Figure 110-3.
4. The weld will build a shelf at the bottom of the joint which you can build on. Make sure when you weave the torch that you make sure the arc reaches the root of the joint to ensure good root penetration. Hesitate on the sides to fill in the weld and prevent undercut. Increase your travel speed when going from side to side to prevent excessive buildup, which would result in a very convex bead.
5. Complete the first pass, keeping the fillet size to as close to ¼ in. as possible.
6. Cool the plate thoroughly and deposit a second pass. Keep the second pass fillet size to ⅜ in. (See Figure 110-4.)
7. Weld the second side of the T-bar assembly using the same technique as the first side.

252 FLUX-CORED ARC WELDING PROCESS

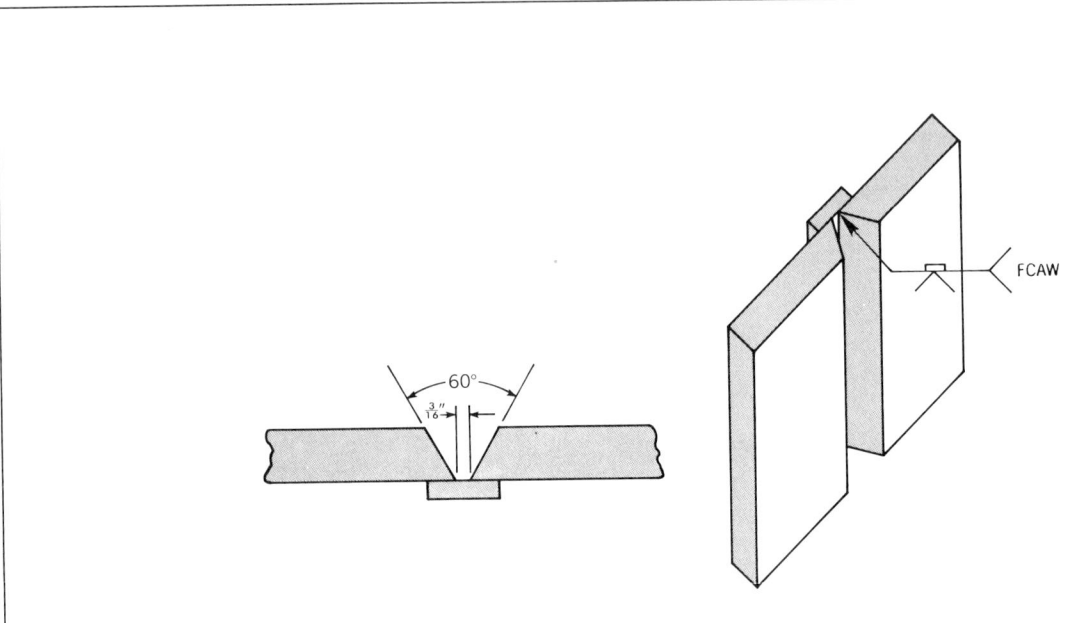

FIGURE 111-1

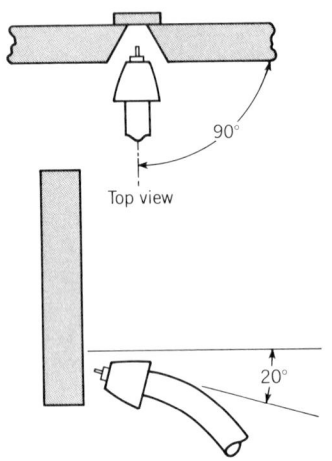

FIGURE 111-2

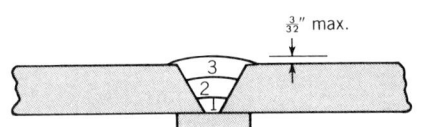

FIGURE 111-3

NAME	Grooves	POSITION	Vertical	
ELECTRODE E-71T-1	DIAMETER .045" Gas flow 30 to 40 CFH carbon dioxide	AMPERAGE 170-190	VOLTAGE 21-23	
MATERIAL	1 pc. ¼" × 1" × 6" carbon steel 2 pcs. ⅜" × 3" × 6" carbon steel with a 30-degree bevel on one 6" side			
PASSES Multiple	BEAD Weave	TIME (SEE INSTRUCTOR)		

Procedure 111

V-Groove Butt Joint, 3G Position with Backup Bar

OBJECTIVE

Upon completion of this lesson you should be able to weld vertical up grooves.

Text Reference: Section VI, Lesson 4C; page 423.

PROCEDURE

1. Adjust the power supply and wire feeder to obtain 21 to 23 volts and 170 to 190 amperes, gas flow 30 to 40 CFH. Stay on the low side of the range.
2. Clean the pieces to be joined tack them together and position them as shown in Figure 111-1. Thoroughly tack the backing bar to the plates.
3. Begin at the bottom of the groove using the gun positions shown in Figure 111-2. Build a shelf for the weld and using a slight weave motion continue the weld up the joint. Make sure the weave is wide enough to cause the arc to melt the root of the joint ensuring good penetration and fusion.
4. After completing the root pass, deposit two more layers using a single-bead weave to produce each layer. (See Figure 111-3.) Increase the width of the weave for each layer. The second layer should be just under flush. This will allow you to reinforce the joint properly. Hesitate on each side when weaving to fill in the weld and prevent undercut. Cool the plate between each layer.

254 FLUX-CORED ARC WELDING PROCESS

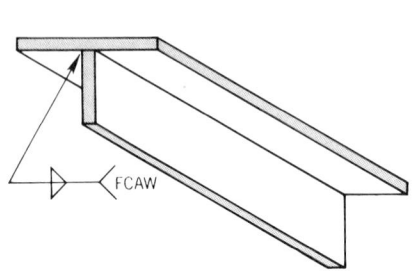

FIGURE 112-1

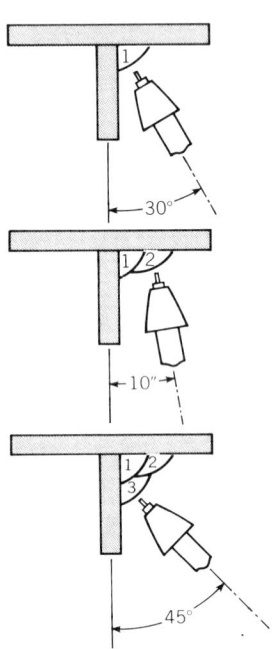

FIGURE 112-2

NAME	Fillets	POSITION	Overhead
ELECTRODE E-71T-1	DIAMETER .045" Gas flow 30 to 40 CFH carbon dioxide	AMPERAGE 170-190	VOLTAGE 21-23
MATERIAL	1 pc. ¼" × 1½" × 8" carbon steel 1 pc. ¼" × 3" × 8" carbon steel		
PASSES Multiple	BEAD String	TIME (SEE INSTRUCTOR)	

Procedure 112

T-Joint Fillet, 4F Position

OBJECTIVE

Upon completion of this lesson you should be able to weld overhead fillets.

Text Reference: Section VI, Lesson 4D; page 424.

PROCEDURE

1. Adjust the power supply and wire feeder to obtain 21 to 23 volts and 170 to 190 amperes, gas flow 30 to 40 CFH.
2. Clean the pieces to be welded, tack them together, and position them as shown in Figure 112-1.
3. Use the gun angles shown in Figure 112-2 and the backhand technique to put in the first pass. Slightly weave the gun to prevent lack of fusion along the toe of the weld.
4. Cool the plate, clean it and put in two more passes using the gun angles shown in Figure 112-2.
5. Cool the plate and weld the second side of the T assembly using the backhand technique and the gun angles shown in Figure 112-2.

256 FLUX-CORED ARC WELDING PROCESS

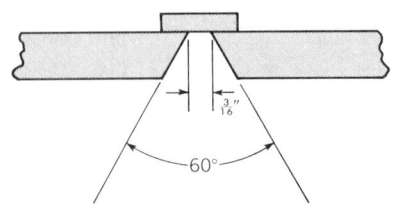

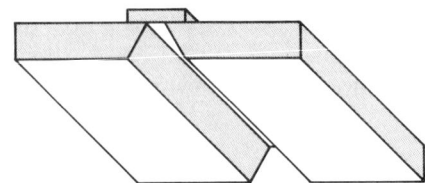

FIGURE 113-1

FIGURE 113-2

NAME	Grooves	POSITION	Overhead	
ELECTRODE E-71T-1	DIAMETER .045" Gas flow 30 to 40 CFH carbon dioxide	AMPERAGE 170–190	VOLTAGE	21–23
MATERIAL	1 pc. ¼" × 1" × 6" carbon steel 2 pcs. ⅜" × 3" × 6" carbon steel with a 30-degree bevel on one 6" side			
PASSES Multiple	BEAD String	TIME (SEE INSTRUCTOR)		

Procedure 113

V-Groove Butt Joint, 4G Position with Backup Bar

OBJECTIVE

Upon completion of this lesson you should be able to weld grooves in the overhead position.

Text Reference: Section VI, Lesson 4D; page 425.

PROCEDURE

1. Adjust the power supply and wire feeder to obtain 21 to 23 volts and 170 to 190 amperes, gas flow 30 to 40 CFH.
2. Clean the pieces to be welded. Tack the pieces together and position them as shown in Figure 113-1.
3. Keep the gun perpendicular to the joint using a slight backhand technique and a weave. Begin at the end of the joint and keep the wire at the leading edge of the puddle. The position of the wire will determine the amount of penetration into the backing strip and shape of the root pass. Keep the wire on the leading edge of the puddle and deep into the puddle.
4. Complete the joint using the bead sequence shown in Figure 113-2. Change the gun angle for each bead to point in the direction of the bead. Weave each bead to ensure good penetration with the side walls of the groove, and other weld beads.